CHELSEA M. RIVERS

HURRICANE IVAN:

THE EXPERIENCE

**DEVASTATION OF, RECOVERY FROM
AND RETROSPECTION ON
A CATEGORY FIVE HURRICANE**

Copyright © 2009 by Chelsea M. Rivers

HURRICANE IVAN: THE EXPERIENCE
by Chelsea M. Rivers

Printed in the United States of America

ISBN 978-1-61579-124-8

All rights reserved solely by the author. The author guarantees all contents are original and do not infringe upon the legal rights of any other person or work. No part of this book may be reproduced in any form without the permission of the author. The views expressed in this book are not necessarily those of the publisher.

www.xulonpress.com

This book is dedicated to my husband,
Paul Rivers, whose belief in me and my
abilities has far exceeded my own expectations
of myself. It is only through his tireless
persistence that this book has seen the light
of day. I am (now) eternally thankful for his
nagging, (what a reversal of roles!), for otherwise
I would not have followed through with this
labor of love (and hate).

We survived this hurricane and the aftermath
together and I would not have wanted to
experience it with anyone else! Paul, you are my
rock and my strength and I count you
and our children, Jordan, Branden & Amelia
as my greatest blessings!

Thank you for believing in me.
Thank you for pushing me.
(Thank you for being Mr. Mom so I could
write in peace!)

"Ivan was a classical, long-lived Cape Verde hurricane that reached Category 5 strength three times on the Saffir-Simpson Hurricane Scale (SSHS). It was also the strongest hurricane on record that far south east of the Lesser Antilles. Ivan caused considerable damage and loss of life as it passed through the Caribbean Sea."

Stacey R. Stewart
National Hurricane Center

Hurricane Watch issued:
Thursday, 9 September 2004 – 0300 hours
Hurricane Warning issued:
Thursday, 9 September 2004 – 2100 hours
Hurricane Warning discontinued:
Monday, 13 September 2004 – 1500 hours

PROLOGUE

The ninth hurricane of 2004 was named Ivan. It was a very real and very deadly storm that claimed 42 lives in the Caribbean. Unfortunately, the Cayman Islands reported one case of death directly associated with the storm, which was attributed to personal negligence. Other deaths that occurred in the following months were due to stress, heat, and the improper use of gas generators to power homes and businesses.

In the aftermath of the hurricane, which lasted 36 long, harrowing hours, it was determined that 95 percent of the homes and other buildings (which generally follow South Florida's stringent building codes) were damaged or destroyed.

Ivan reached Category 5 three times during its lifespan, the highest possible rating that is assigned to hurricanes, where wind speed can reach well over 75 miles per hour and gusts can exceed 200 miles per hour.

The stories depicted in this book are based on actual events that happened to real people. Some creative license has been taken by this author, but overall the details within are true and accurate accounts.

Of the three islands that make up the Cayman Islands, the largest of them (Grand Cayman) took the worst beating. To a much lesser degree, Cayman Brac and Little Cayman sustained some damage, but were quickly returned to normal within the month of September 2004. It took less than one year for Grand Cayman to be restored to a state of operation deemed suitable for the return of tourists and outsiders.

For the duration of 2004, Grand Cayman was on lock-down to visitors and international news media. Water and electricity remained out in some areas for this entire period of time. Only Caymanians, residents, or people with physical property were allowed to enter as the country came to grips with the monumental task of cleaning up and moving on.

The governor of the Cayman Islands was ultimately forced to declare a state of emergency.

The people of this wonderful country did what came naturally; they pulled together in a unified comportment and neighbor helped neighbor until life resumed the familiar pattern of daily living.

This hurricane forever changed the lives of those who chose to ride it out. It connected people who otherwise would have never been brought together. It opened the minds of those with their noses too close to the grindstone. It set perspectives in a different direction and from a new angle.

The survivors of Hurricane Ivan now share an experience that will travel with them for the rest of their lives, like a scar that never fades. It is a scar that is worn with pride, as anyone who lived through it will proudly tell you that they never left – never abandoned the ship. Locals and foreigner survivors alike are bound by the common thread that has woven them together in the fabric of a historical event that affected them in different ways.

It was both the worst and best thing to happen to Cayman. Five years later, it is still the subject of conversation everywhere you go, and the details and feelings are as fresh in the minds of the tellers as if it were only yesterday.

The Cayman Islands are a blessed trio built on strong Christian morals and principles. The protection of the Almighty was never so obvious as in September 2004, when their inevitable collision course with a deadly storm that should have entirely decimated all life and matter bore the mark of divine intervention, as all inhabitants emerged relatively unscathed and alive.

May this book serve to remind our people of their blessings and continue to force our eyes heavenward with gratitude for the day that our lives and country were spared from Hurricane Ivan. Let us never forget, nor take what we have for granted. Let us remain prayerful and devoted to ensuring our islands are governed with great fear and respect for our Heavenly Father.

ONE

September 11th, 2004
0600 hours
Ship call sign: WDA406
Hurricane Report: Latitude 15.9 Longitude 77.5
Approximately 375 miles SE of Grand Cayman
Category: 5

"Well folks, it's hurricane season again and you know what that means…time to batten down and fight those other shoppers for the last can of corned beef!" The weatherman on the local television station was standing in front of an enlarged map of the Caribbean Sea, which was enshrouded in an enormous white globular cloud. The swirl was in an electronic looping sequence that caused it to rotate in an upward moving northwestern direction, and then suddenly drop back to its original position and begin again. Its edges reached from the northern tip of Honduras to the southern coast of Cuba.

"Yes folks, this is serious. Hurricane Ivan is heading straight for us and if you haven't already got your house in order, you better be packing a bag and getting to a shelter." He smiled and let out a chuckle as he clicked a button on his hand-held device to change the picture on the screen behind him. The image zoomed in to focus on the Cayman Islands. The white cloud was gone and the familiar red symbol of fused open and closed quotation marks depicting a hurricane now sat in the middle of the blue ocean, appearing very harmless. It was a mere dot in comparison to the wide bodies of land that hemmed it into the Caribbean Sea.

"Here, ladies and gentlemen, you can see the direction of the eye." He made a single pumping gesture of his hand, and the screen mysteriously altered to include a yellow funnel-shaped path stemming from the eye and spreading out to encompass the three islands.

"Now remember, this is only based on predictions, anything can happen. For now, the national weather service is predicting that it will pass right over us and being a Category 5, well, let's just say expect a lot of flooding and roof damage." The smile was gone and replaced by a look of worry. He was generally a jovial weatherman, come rain or shine he delivered the news with a wink and laugh.

But that was not the case now. He knew what was coming; he had seen it in his research of historical weather systems and damage that hurricanes alone can cause. He knew that they traveled in different categories, on an ascending scale of one to five. Wind speeds determined how devastating they could be, and gauged how it ranked on the Saffir-Simpson Hurricane Scale.

HURRICANE IVAN: THE EXPERIENCE

"A category five is deemed the most dangerous of all, with wind speeds that exceed 75mph and gusts that can reach well into the two hundreds. A hurricane of that intensity can surely cause roofs to separate from their houses and roll cars down streets," he announced in an uncharacteristically serious tone.

He was looking straight through the camera, into the eyes of every person who watched him now on their television sets. He wanted to tell them that it would almost certainly leave behind families with lost loved ones. Ripped from their comfort zones, some would mistakenly try to run into the storm in search of shelter. Others would never even have the chance, when their world was crushed by the power of the storm. A hurricane of this force is not a mere storm – it is a weapon out of control, indiscriminate and ruthless.

He knew that throughout history, there have been several category fivers. In each case, the report of death and destruction was extensive. Billions of dollars in damage to home and property can never compare to the thousands of lives lost, however, the two go hand-in-hand when a weather system of that magnitude bears down on your country. Worse if you happen to reside on an island, where all around is the expansive sea that can be coerced so easily by the gentle whisper of wind, causing shivers to ripple on its liquid skin. Imagine the effects of persistent gusts that buffet the once still, silent waters, driving it into frenzied motion of wave upon wave, building to tidal wave strength. Waters that invited you into their cool, refreshing embrace, offering you a sense of comfort and relaxation, could be so quickly transformed into a monster of unspeakable proportions; seeking every nook and cranny to fill with its airless hug.

The weatherman shivered as he tried to focus on what he had to tell the public. Images of news reels from other countries in Ivan's path flashed before his mind's eye as he fought to regain his composure and do his job.

"I am sure each of you picked up your hurricane tracking maps at the local hardware store or supermarket, so get ready to plot the newest coordinates." Again the screen behind him changed to reveal a photo of a bruised sky. It was the underside of the circular mass of clouds that had been shown at the start of his segment. A series of numbers appeared in large white letters.

"Hurricanes are tracked by coordinates plotted on a graph provided by storm planes, called Hurricanes Hunters, which fly into the heart of the system to gauge wind speeds and direction among other things. The globe is plotted out with every country assigned a set of grid digits according to their placement on the surface of the Earth." He now transformed into a professor of weather – an official meteorologist. He pushed his glasses up the bridge of his nose as he laid out the facts. Sweat stains appeared on his blue shirt under his armpits, and slowly spread forward.

"As the hurricane moves forward, the location is updated frequently to provide a visual line of where it has come from and estimate where it will go. The Tropical Prediction Center or 'TPC' can use models from years of research, based on the patterns of every hurricane in the last hundred years. Although these storms do not travel in a ruler straight line, their wobbling path can be predicted with impressive accuracy. Countries in the path of a hurricane begin getting warnings in advance to ensure proper precautions are taken and that evacuation procedures can be initiated if need be."

He knew that he might have lost half of his audience to a more interesting program, but he felt it was his duty to break it down for everyone watching. There was very little that he could do physically to assist in the island-wide preparation for this enormous weather system, but he also knew that knowledge was power and right now, he was a very powerful man to pay attention to.

"Now, if you have been asked to evacuate your area, you will need to consider moving to the nearest shelter. Speaking of shelters, the beautiful Melissa is here to remind you once again where they are located." The image switched to the news anchorwoman, who began to list out the various locations and announce the shelters that were full and no longer accepting new arrivals.

Now off-camera, the weatherman removed a handkerchief from his pocket and dabbed the sweat from his forehead and upper lip. The cameraman wagged a finger at him, but he pointed up at the bright stage lights and mouthed, "They're too hot." This explanation seemed to satisfy the teaser, who turned his attention back to his equipment. The weatherman stepped down from his platform in front of the green screen that had earlier displayed the true object of his profuse perspiration. He knew full well that the studio lighting did not cause the droplets of body fluid that purged from his pores. This reaction was a direct manifestation of his fear of Ivan and the knowledge that this was no harmless storm. This was a powerful machine procured by nature to teach these islands a lesson.

TWO

Mama DeDe was worried. She had seen the news, and knew that what was to come was like no other storm this island had seen before. She was not the type of person to sit and fret. She was a woman of action, and at the moment this situation called for some seriously fast action. She was going to have to be a superwoman to get through this. There was, of course, no better person for the job. The first order of business was to stock up. In order to survive the aftermath, you needed to have enough supplies on hand *after* the storm to ensure that you would be alive in the weeks to come. Just preparing for the storm itself was good, but a superwoman needed to know that whatever came after was of no surprise and that she would be well equipped to face the challenge. She pushed the knot of anxiety deep within and proceeded to ready herself for a trip to town in pursuit of each item on her carefully thought out list.

The drive to the store was as uneventful as any other. In fact, the day was such a beautiful one that it was difficult to imagine that anything so menacing could possibly be threatening the shores of such a paradisiacal island. The sky stretched before her, peeking out behind buildings and almost shoving them aside to hold her attention, as she casually passed row after row of tourist accommodations. The blueness of the atmosphere was a perfect blend of dark and light. Fluffy white clouds carelessly bounced along in their whipped state of perfection. A painter's brush would be hard pressed to depict a more serene heaven.

The sun was slowly burning its way across the last of its 90 degrees of the morning quadrant in its rotation. Blazing brightly, it basked in its own glory as it crisped the layers of skin on the bodies of worshippers laid out on the hot, white sands, who spun as if on rotisseries, in search of the perfect tan.

Mama DeDe was not a tanner. She had been genetically blessed with the eternal pigmentation that most Northerners were dying for. Her golden brown complexion gave her an everlasting glow of health and like her mother; she enjoyed an ageless beauty afforded to many Caribbean women. It was difficult to determine that she was already halfway past fifty, for she bore no wrinkles or telltale signs, and always carried herself with grace and posture. She did not rely on miracle creams and anti-aging concoctions to maintain her youth. She simply chose not to engage in a lifestyle that would be considered foolhardy and reckless. This included spending too much time in the sun; a complete contradiction of the perceived rule of island living.

Both hands on the wheel, she checked her rearview mirror often. A good driver always knew what was before and behind them. A careful driver was sure to keep the required distance between vehicles, depending on the speed at which one traveled. She navigated her large vehicle along the well-traveled road with much caution and attention.

Flicking her indicator on, she signaled her turn into the parking lot of the supermarket in search of a space close to the building. As she edged around the turn into the first row, she spotted reverse lights on a van that was parked just outside the main doors.

"C'mon baby," she breathed as she sped up to take the spot before anyone else could. The driver slowly backed out, opening the gap for her to pull right in. She slowed to a stop just as her tire bumped the bright yellow parking block.

What luck! she thought, as she shifted her gears into park and switched off the engine. The vehicle made several pings and hisses as she gathered up her purse and car keys. It was a big machine for a woman, not quite a truck and not a car by any stretch of the imagination. It certainly earned its name, Avalanche, just on size alone. She loved it and would not trade it for any other, even on days when a good parking space was hard to fit into. It was just a year or so old, and still ran like a dream. The interior was tan leather, which complemented the deep blue exterior tastefully. It was no doubt the best automobile she had ever owned.

Standing outside, a cool breeze caught her in the crosswalk between the parked cars. She reached up to hold down the wide-brimmed, sun-repelling hat she wore and smiled at her wise choice to don slacks rather than a skirt. With her free hand, she clicked a button on her remote to engage the locks on her vehicle and waited momentarily for the chirping response, which signaled the action was complete, before she walked away in search of a cart.

The blast of cooled air buffeted her face as the motion-sensing sliding glass doors opened with a swooshing sound. Mama DeDe glanced at the first items on her emergency supplies list and steered her cart toward the aisles of interest.

Water was priority, and then the dry foods. If electricity was cut and restoration delayed for an extended period of time, fridges would be useless; therefore buying ice cream would just not work in this particular situation.

"Hey DeDe, that you?" a voice from behind called out to her. She turned to see an old friend approaching her, shopping cart piled high with cans, bottles and flashlights.

"Yes chil', it's me. Trying to stock up, ya nah!" DeDe replied as she reached out for a hug. The two women embraced warmly before releasing and returning to their shopping carts, which they moved closer to the shelving to open the aisle to the traffic of other shoppers.

"Darlin' I get caught out every year by waiting until the last minute to stock up," her friend sighed. "This year I swore I was going to do my shopping early, at the beginning of the season, but look at me – already three and half months gone and I am jus' getting 'round to it. If we didn't have this hurricane on the way I would probably still have done nothing!

I don't even know if I can find any plywood now!"

Mama DeDe laughed. She had already bought most of the big ticket items like lanterns and flashlights in June, when the season started. Her house was equipped with large aluminum awnings which were characteristic of many old-fashioned Caymanian homes. As they remained fastened in place after their initial installation, she enjoyed the shade they provided year-round. Above each window the hinged, corrugated metal awnings were tightly secured, allowing the home-owner to simply lower them over the window during bad weather with little effort. Besides their protection function, they were a beautifully quaint adornment which she insisted on purchasing when she built her home, so thankfully plywood was not an item on her checklist. She felt sorry for those who had to buy the wood because price gouging usually ensured that it was one of the most expensive purchases one would make to secure one's home.

"I know what you mean," Mama DeDe said. "Some years we have nothing to watch out for and you end up left with all this stuff."

"That's right!" her friend agreed. "Last year I had so much canned food I ended up donating it to some relief drive for Haiti or Cuba or somewhere. It wasn't a total waste in the end, but it's still frustrating buying all of it and nothing happens."

The two chitchatted a few more moments before parting company in search of the remaining items on their lists.

"Be sure to get extra bleach," Mama DeDe called

back to her. "I think it's on sale today." She waved farewell as she pushed her cart around the end of the aisle and towards the bottled water.

By Wednesday, Mama DeDe had completed the shopping needed to stock up. She was sure to catch the deals and sales at every store. The benefit of being retired afforded one the chance to scout out the best prices and be the first in line when a sale was announced. She loved to shop, but there was no joy in standing in line with an entire island of people looking for the same items: candles, batteries, canned foods, dry milk, and flashlights, among other things. Businesses knew what to expect and therefore during hurricane season, they increased their shipments with these types of products.

At 3a.m. on Thursday, a Hurricane Watch was declared on the Cayman Islands. Mama DeDe awoke to the news and immediately her mind was turning with ideas of what she would need to do to properly prepare. She knew that she lived in a low-lying area and that the sea was about a mile to the east and west of her. She decided that sandbags would be needed to keep the water out of the bottom level of her home, should the sea converge on her property or the rains dump water so heavily that flooding would be inevitable. This required canvas bags and fresh beach sand, not a problem; she would attend to this at once and move on to the next requirement. She was, after all, a woman of action and therefore could handle most any problem on the spur of the moment.

Creating sandbags was no small task. It required a working knowledge of sewing and a creative mind.

She purchased the canvas from a cloth store, and after careful measurement, cut it into the required lengths to yield several very attractive bags. With the help of her sister Nay-Nay, they were able to scoop buckets of clean, white beach sand from the shores and fill the bags, tying them off with nylon cord. Proud of her handiwork, Mama DeDe set out to complete the remaining items on her to-do list with careful consideration. She could not rush this, otherwise she would forget something of vital importance. Every step must be planned out meticulously from top to bottom. There would be several people at her house and under her responsibility; she must pay close attention to their needs and think everything through in order to have a complete plan of action.

Mama DeDe needed to arrange to have her shutters lowered over her windows. Having a two-story home made this a bit tricky, but she knew the person she called would be able to handle this. After making the arrangements, she surveyed her home. There was so much to be done and so little time. 9 p.m. brought news that the Watch had been upgraded to a Hurricane Warning. It was inevitable. The only question that remained was just how deadly would it get.

THREE

Fiona knew hurricanes could mean trouble, yet there was so much excitement about this one. The islands had been through so many false alarms that it was difficult to get excited anymore when the weather station declared a watch was on. The usual activities always accompanied such dreaded announcements; last-minute shoppers trying to stock up on the items that might be needed. Hustle and bustle everywhere.

"I bought all of dis stuff for the hurricane. I really hope that we see some action dis time!" she announced with a laugh, "I's getting tired of being all prepared for dese t'ings and fighting da lines at da stores is not my idea of fun on a Sat'day." She carefully extracted a carton of eggs from a sack and deposited it in the refrigerator, her slender fingers catching on a jar of mayonnaise and almost causing it to crash to the ground. Quick reflexes enabled her to right it before catastrophe struck, and she swiftly turned to the shopping bag once more.

"Be careful what you wish for, you just might get it," her husband chided her softly. "Besides, if you had stocked up a week ago, when we first saw this one coming, you would have been able to relax today." They had only been married a year and their union had produced a son who was now three months old. Little Evan lay asleep in his car seat by the front door as Fiona brought in the bags of supplies from the car. She was hopeful that he would catch some cooler air closer to the door.

"Oh Dan, you know I's just kidding. I would never wish anyt'ing like that to happen to us. Besides, after da las' hurricane passed tru' here without so much as a breeze, I am certain that dis one will barely even trouble us either."

"I hope you're right, honey. But from the looks of it on the weather station, this one might be a little nasty." He passed her, encircling her petite waist briefly with his large arms and pausing to gently peck her on the neck as he made his way to the back door to finish putting the shutters on the remaining windows of the house. Dan was almost twice her age, but they felt that the years between them were insignificant. They reached each other on a level that no one else could. For them, their love was immeasurable and was all that mattered. She was a tiny thing compared to him. Barely five-and-a-half feet tall, she weighed no more than a hundred pounds, but being the youngest girl in a house of several big brothers gave her a survivalist mode that intrigued Dan.

"I leave you to the easy work of unpacking. I need to finish the manly work of securing the house before Ivan hits."

"Don't forget to trim da branches on da fruit trees.

Da last t'ing I needs is to have da limbs crashing tru' da roof." Fiona glanced at Evan, who was stirring in his temporary bed. His mouth was working on an imaginary bottle that he was busy draining of its airy milk. His olive skin shone in the sunlight that streamed in through the open door; ample feedings were evident in the tautness of skin over the plump arms and legs. She paused as she watched for signs of his awakening, to determine if she would have to abandon her chore at hand in favor of quelling his hunger-filled cries. After a few twitches, Evan lulled back into the land of the Sandman.

Satisfied that she would not be drawn away from her groceries, she settled back into unpacking and preparing her house for the many guests she would be hosting for the next day or so. Her home would soon be filled with family seeking shelter, and she knew that if things turned out for the worst, she might be pulling an all-nighter.

By Saturday, everything was going according to plan and her checklist was almost complete. Driving along the Seven Mile Beach, Mama DeDe scanned the buildings as they seemed to rush past her. Would these still be here tomorrow? The colors were radiant in the midday light. Glass from the shop windows winked at her as the sunlight caught them at certain angles. People casually walked the long sidewalks that hugged the road, chatting and laughing. They seemed so unconcerned about the future. The government had begun to evacuate tourists several days ago in preparation for the worst. What could these people possibly want to stay for?

The liquid crystal display on her dashboard clock showed ten minutes before noon as she drove to pick up her friend Lacy and Lacy's eight-year-old daughter, Chloe. They would be riding out the storm with her. The mood was more festive than worried. It was like a sleepover for adults.

"Is that all you plan to bring?" Mama DeDe teased, as Lacy tossed several overstuffed bags into the back of the Avalanche.

"Girl, I might as well be moving in with you. I don't think I have anything left in my closets." Lacy laughed as she helped Chloe into the backseat and settled herself into the front. She had been one of Mama DeDe's best friends for what seemed like an eternity. She was always available and willing to help out, whatever the request. Mama DeDe could not think of a better person to wait out a storm with.

"DeDe, do you think that this one is going to be as bad as the '32 storm?" Lacy asked, remembering the stories she had heard about since childhood.

"Ah child, that storm was bad, but I really believe that Ivan will put it right out of our minds for good."

The hurricane of 1932 was the most disastrous storm in the recorded history of the Cayman Islands. Many had died, and each generation was reminded of the devastation that nature could inflict if one was not prepared.

"Well, I hope you got enough food and water, because I am not leaving until the coast is clear!" Lacy stated.

Back at Mama DeDe's home, they proceeded to wrap as much of Mama DeDe's belongings in plastic as seemed reasonable. The television set, clothes, all of her treasured pictures and many other items were carefully packed in plastic bags. It seemed an overly cautious exercise, but she had heard all of the horror stories about the 1932 hurricane and was taking no chances. Besides, she was retired now and she had plenty of time to put everything back as it was after the storm had passed. Who knew, maybe she would use this opportunity to redecorate altogether.

In the event of something terrible happening, she did not want to lose anything within her home. Everything was so precious to her – especially her photos. The pictures of her twin grandsons graced the walls in almost every room. They were her pride and joy, and she could not bear to see these mementos of times they had shared destroyed. Here, they were celebrating a birthday, all smiles, eyes twinkling with juvenile impishness. There, they were posing in some costume that she or Nay-Nay had made for them. Pirates in this picture; clowns in that one. Her heart swelled with pride. They were such fine young men. At seven they were mature boys, full of so much energy and spunk that it plumed around each photograph of them like the fumes off of fuel. You could feel the joy from the photos pouring out through the paper into the air. It heartened you to walk through these rooms and see them laughing back at you. There was so much love in this house. She eyed a small toy that they had played with on their last visit and neatly placed it in their toy box, then set out to remove every photograph from the walls and tables.

FOUR

The wind was pleasantly cool and refreshing. A wonderful distraction from the unbearably sweltering heat that had settled like a thick molasses on these three land masses known as the Cayman Islands, covering everything in a blanket of suffocating stillness. Summer can be brutal in the Caribbean. Even the breezes that floated in from the ocean were more teasing than cooling. Nature's cruel joke. Another method of sending a sane person over the edge. On this night in September, the high anticipation in the air chilled to the bone. This wind was welcome only for the brief interlude that it provided from the fire of summer. Not as the prelude for the enormous storm that hovered on the outskirts of these tiny islands, eyeing his prey with lip-smacking hunger. Ivan was coming – Ivan the Terrible – and he was intent on devouring these three islands with an appetite as voracious as any malnourished lone wolf, ribs protruding, flanks sunken from thirst, seeking a meal long overdue.

They sat on the porch of a neighbour, this small family of four. Des, the head of the household, was a quiet and unassuming man of more than 6 feet in height. His wife, Michelle, easily shorter by a full foot, was irrefutably the backbone of the family, yet she always knew when to gracefully yield to his authority in their home. They were blessed with the arrival of twin boys two years into their marriage. Dylan and Demitri were now handsome young boys who enjoyed an equal mix of their mother's looks and their father's temperament.

They intended to ride out the storm there at the home of their good friend, Jackie. Not because they were uncertain that their own home was able to handle the force of the winds, but simply to have company, conversation, perhaps enjoy a reprieve from the nervousness of the unknown. Sitting there, the tone was calm and relaxed, but there was an underlying voice of uneasiness, a sense of urgency to flee; leave before it was too late. As the night pressed on, uneventful, save the constancy of the wind, they sought to keep their minds occupied with the talk of work, their children, anything that would not involve discussions of the weather.

As they sat there in the glow of the streetlight, the rain began to fall in lazy sheets, dotting the bone-dry pavement of the driveway. The plants and trees jumped to attention, with the drops dancing on their surfaces causing their limbs and leaves to quiver and bow. Des and Michelle corralled their boys and pulled closer to the walls of the building to avoid the droplets that sought to possess bare arms and legs with their icy chill. Within moments, the wind that had persistently pushed through the empty streets with driven determination, like a doomed spirit on

HURRICANE IVAN: THE EXPERIENCE

the prowl for a soul, suddenly raised its howl to a new level, and threw dust and pellets of rain at all who were in its path. The conversation ended abruptly and they raced inside to take shelter from the storm. He had arrived.

"Whew! Looks like the time has come!" Des announced excitedly. He quickly did a head count to ensure all were accounted for. He was a strikingly tall man of slender build. Broad shoulders carried an endless supply of hugs and breadth to comfortably nestle the sleeping head of either of his twin sons, as they were often called to do. His face was youthful, disguising his true age. It was difficult to believe that he was approaching his 35th birthday. Des was a gentle spirit who found pleasure in the simple things in life. Now in his tenth year of marital commitment to Michelle, he was fine wine in an open bottle, breathing.

"Okay boys, we are at the point of no return now! Ivan is almost here." There was a playful tone to his words, but inside he was suffering from the tautness of that vice that clamps your chest tightly, unrelentingly constricting with every exhalation. He was so very wary of the hours to come. Looking at his babies, he could not help but feel that surge of emotion well up in the back of his throat at the prospect of the unknown.

Michelle had been in a panic for the better part of the week. She had been glued to the weather channel and the hurricane websites, and insisted on tracking them herself. She also compared them to other hurricanes that had come their way, and was resolute in her belief that this one was different; it was going to be worse, much worse. All week she had been nagging him to put the shutters down, to clean up the junk that had been accumulating around the house and properly store everything in the shed. All week he had been promising to do just that, but just never got around to it. Man, would he hate it if she were right on this one.

Now on the very evening that the hurricane was to hit, Des was happy that she *had* nagged him. She had of course done much of the work herself, with the help of their sons, Dylan and Demitri and their nanny, Gela. Eventually her persistence struck a chord and he realized that she might be right.

Now they settled inside with a nervous laugh. The expectancy of the birth of Ivan was short-lived, only one week's notice. Enough time to prepare materially, but until the day came, until the very moment had arrived that you knew there was no escape, that your fate was sealed at the instant that you *felt* his presence, there was never enough time to prepare mentally. The fear was ever present, its cold, bony fingers icily running down their spines, rapping at that inner door of their minds that only opens when Fear knocks. Opens to admit thoughts like visitors, thoughts that pour through in ceaseless flood. *Will you die? How? What would it be like to drown? Will I feel it? Will the roof blow off? Will I be sucked out? What about the children? Oh my God, I can't do this! How do I get out of this? Is it too late to leave? What was I thinking?*

The knots in your stomach that accompany fear

are as ruthless as bad Chinese food, tearing through your gut like shrapnel, ripping you from the inside. That churning low in your bowels that becomes stronger the more rampant your emotions become. It is this fear that can cause your death; that can paralyze you at the very moment that you need to be running, fleeing. The "deer in the headlights" syndrome that has brought about the end of so many lives by the inability to get your body to react in motion with the screaming in your head. When your feet remain fixed to the spot, unable to move while your brain cycles through every possible route of escape.

Yet, this fear can also cause you to perform feats of amazing strength. Triggering your adrenaline to flow like molten lava, hot through your veins, releasing power of unspeakable measure. Mothers able to fend off vicious animals to protect their young. Men capable of lifting objects that would have otherwise been unthinkable to heft, in order to rescue trapped loved ones.

What do you do in fear?

Fiona welcomed her guests into her home with a sense of pride. She had built the house upon the proverbial rock. Everyone was here to seek shelter and enjoy her company. She was thrilled that they had chosen her home.

Her mother had arrived with her grandmother, her brother with his wife and five children, too. Dan's mother sat on the sofa, watching in amusement as the youthful guests ran back and forth in front of her. The children sensed the excitement and were busily playing tag around the house. It was so noisy, almost like Christmas Day. She really didn't mind, as she loved being surrounded by family. Dan was still busying himself with the last-minute preparations.

Fiona made another tour around the house before her mother called her in to start dinner. They were to have a feast tonight. She wanted to make sure that there was enough food for everyone. Her other brother would arrive shortly, once he finished securing his own home.

"Fiona, you better mek dem kids settle down now," her mother chided as she got up and joined Fiona in the kitchen.

Fiona knew her mother's nerves were rattled. She was accustomed to having her grandchildren running around her ankles, but the circumstances were so different than a casual Saturday evening. Fear of the unexpected impregnated the air with an almost palpable odor.

"Oh, Mummy, leave dem alone," Fiona almost snapped, but caught her tone in time. "Dey are havin' fun and keepin' out of da way. Go lay down an' relax. I can get Julia to help me get dinner together," she said, referring to her brother's wife. She waited for her mother's retort, but was relieved when the older woman dropped her shoulders in acquiescence and slowly walked towards the bedrooms.

Julia stepped to where her mother-in-law had stood and winked. "Okay, Fi, whatcha need me to do?"

"Can you put da potatoes on? I needs to cut the veggies." They each silently took to their individual tasks, whilst simultaneously watching the brood of youngsters that continued to race around the house.

Dan returned from outside, sweat dripping from his head and face in skinny veins of liquid. He looked exhausted and did nothing to hide it from Fiona. She quickly ripped two paper towels from the spool on the countertop and walked to meet him as he entered the kitchen, her hands stretched outward and upward.

Instinctively, he lowered his head to meet her hands, and she gently wiped the manly perspiration from his shiny crest. She had never seen him in his younger days, when he allowed his hair to grow out. Of course, she saw pictures, but it was hard to imagine what he would look like now, had he not insisted on keeping his crown shaved.

Her hands patted down the sides of his face, catching a stray bead of sweat that threatened to bungee jump from the tip of his nose. He smiled as she leaned in towards him, planting a delicate kiss where she had just rescued the errant pore runaway.

"Well, the shutters are all pulled across, the trees are all trimmed and every piece of junk that couldn't be tied down is inside. Let me tell you, I thought I worked hard, but your brother Ernest is a powerhouse." Dan leaned against the countertop to support his weary body. He was numb, but satisfied that he had done everything he could to ensure that this home was secured. He and Fiona had chosen the more expensive accordion shutters for their windows. They were permanent fixtures which could be drawn across the windows for protection from weather, prying eyes or intruders. With the twist of a lever, they unlocked and could be pushed back into the window recesses in seconds. He was delighted that they had chosen to pay a little more for this added convenience.

"If that storm gets in here, it's not because we didn't try our best to keep him out," he laughed.

Ernest's entry into the house was well received by the children, who chanted, "Daddy! Daddy!" before he was fully inside the threshold. They circled around his legs, each trying to be the one lucky enough to be snared and hoisted into the air for a moment as a helicopter. Even in his obvious state of fatigue, he lifted each of the younger ones for a minute or two, swirling them over his head while the others imitated the whapping sounds of blades on a chopper.

Julia looked on in admiration as her husband attended to their children with interest.

Dan and Fiona watched, as Ernest collapsed onto the floor, allowing his children to climb and jump all over him. He seemed to relish the abuse, as all that could be heard was his laughter commingled with the delighted squeals from each of the five little people he played with.

"Go bathe," Fiona whispered. "Dinner will be ready in about a half hour."

Dan smiled tiredly as he left her to finish her work. She returned to the cutting board where the carrots still awaited their date with the guillotine. Julia was adding the last of the potatoes to the boiling water when the rain began to beat heavily on the shutter that covered the kitchen window, heralding the arrival of Ivan with royal fanfare. The sound was so sudden, each of them had to pause for a moment to absorb the meaning. Hundreds of bony fingers of water thrummed on the aluminum covering, testing its strength with a deafening roar. The wind joined in with powerful fists that tried to find weaknesses in the security of each awning, pounding hard in anger. Deep gusts pushed under the metal frames, lifting and shaking with incredible force.

Julia shot a look of concern towards Ernest, who lay frozen on his back, one child suspended overhead. The others, like wax figures, stood motionless, eyes drawn towards the pervasive sounds coming from every window around them.

A loud and persistent knock at the door startled them, all heads simultaneously jerked in that direction. Ernest slowly lowered the helicopter child, whose outstretched hands had quickly gripped his father's neck and would not let go. He arose from the floor, child in arms, and moved cautiously towards the door. The banging came again, quicker and louder, but before Ernest could put his free hand on the knob to release the door from the frame, it swung inwards with force and abruptness, barely leaving him time to withdraw his hand to avoid contact.

Fiona's other brother, Ray, burst across the threshold, arriving just in time to complete her guest list. The blustery wind and rain needles pushed in from behind him, whipping the curtains into a wild dance and soaking the floor. He stood there, thoroughly drenched and looking very miserable.

"Unna couldn't open the door for me?" he asked,

reaching behind him to press the door back into its rightful pocketed frame. "And where the heck did that rain come from all of a sudden? I was just getting my t'ings out of the car and it was completely calm and dry, but before I could even close the door, I was almost blown away!"

Ray wrestled with the door as the winds took hold and pushed back, holding the gateway open long enough for more enemy rain pellets to infiltrate this impervious abode. Ernest reached around him, son still clinging to his neck, and threw his shoulder into the wooden barricade with brute force. The two men succeeded in slamming the door hard into place, sealing off the attack that continued to rage outside.

For a time longer than a standard minute, no one moved or spoke, as they again listened to the heavy drumming and wind rockets that levied against the house. Only when Dan returned from his shower, with clean towels for Ray and some used ones for the floor, did they slowly break out of their spell and resume the various activities that had occupied them before everything had turned upside down. Ernest took the old towels from Dan and distributed them to his older offspring, motioning for them to clean up the mess before someone got hurt.

FIVE

It was late Saturday evening and the hurricane was now bearing down on Grand Cayman. Most businesses had closed early to allow employees time to get home and take care of their personal affairs and last-minute preparations. Mama DeDe was satisfied that she had prepared as much as she could.

Her elderly mother lived with her now. In her advanced age, she had begun to forget things, so it was thought to be in her best interests that she take up residence with one of her four daughters, to be properly looked after. Mama DeDe was the youngest and, being retired, was happy to take over the care of the woman who had done so much for her.

Beatty was a well-known stalwart in her home district, so it had not been an easy decision to remove her from there. Sadly, aging was unkind and ruthless, and maliciously stripped the elderly of their ability to function and survive through a regular day. The family had decided earlier in the year that it was simply too dangerous to leave her to her own devices any longer, and Mama DeDe stepped forward to accept responsibility for her care.

Now more than ever, Mama DeDe was glad she

did. She ensured that her mother was comfortable and they settled down to wait for the event to happen.

She lived in an affluent neighborhood, which was pleasantly quiet. It was far enough from the heavily trafficked West Bay Road to enjoy the serenity that she needed, yet only minutes from the popular strip that was home to the heartbeat of the island's entertainment and tourist district. Her location was perfect in her opinion. She was never too far from her daily haunts, and at the same time, happily nestled away from the hustle and bustle. She resided in a two-story, three bedroom home on a tree-lined street that shaded the driveways of many white color professionals.

Mama DeDe had been retired for a number of years but had the acumen to realize that she needed some income to pad her future. As such, she had not long ago converted part of a large storage room downstairs into an efficiency apartment that yielded her a nice monthly paycheck.

Her tenant, Massoud, was a nice gentleman who was respectful and pleasant. He had checked in with her as he got home, having spent the early part of the day finalizing his company's business continuity plan. He was happy to stay on with her there, not really knowing what to expect. He had never been through anything like this, but her confidence that everything was going to be all right bolstered him. She had provided him a list of necessities and even purchased some things for him when she did her own shopping. You could never be too prepared.

As the night began to fall, the last two members of the party arrived. Al had helped her with the shutters earlier in the day. He was a trusted friend to her and someone who would help her out at a moment's notice. Naturally, she had insisted that he come right back to stay with her and her mother during the storm. Al was a fine Jamaican man who had always been the muscle for any task or chore that she could not personally attend to. She was so blessed to have friends like him and Lacy.

Al had brought his nephew, Chris, an extra pair of hands and shoulders, should the night's event take a turn for the worst. Though Al had politely declined her offer initially, his conscience had pricked him greatly at the thought of leaving these ladies to fend for themselves, and he and Chris had decided to take her up on the offer at the last minute.

Mama DeDe led them upstairs, where they all stood in the kitchen as the radio recycled the governor's address to the islands, and each stopped mid-conversation to listen as his voice permeated the rooms in a soft, humble tone.

"May God bless and preserve our island and all our people and visitors. My thoughts and prayers are with you all. Good night." The radio personality came back on with more information about the shelters and advice on how to batten down. Nay-Nay reduced the volume so that it continued to be heard, but allowed them all to chat without interruption.

The preparation and enjoyment of dinner had been interrupted by the intrusion of unwanted guests in the form of leaks. The children were now confined to a bedroom that had thus far been spared; however, bubbles on the ceiling foretold of impending seepages that would mature into full-blown leaks in short order. Fiona and her family had been battling this situation for several hours.

"Dan, can you bring me another bucket for dis room? Another leak jes' sprang up." Fiona was tired, hungry, and wet. All across the house, buckets were catching water from leaks that were mysteriously popping up without warning.

"We don't have any more," Dan called out from the kitchen. "You took the last one for the bathroom leak."

"Well, bring me a pot, den."

"You used the last one of those for the leak in the hallway," he called back. "I found a plastic container. It's not very big, but we can keep dumping it."

"Whatever, jes' bring me sometin'!" she yelled impatiently. She needed some sleep and some dry clothes, but if she stepped away for even a minute, there would be no way to keep the intruding water from taking over.

A loud crack coming from the living room alerted everyone and brought the children out of the bedroom. The sound resembled a firecracker. Fiona raced them back into the room and ran to survey the damage.

She found Dan and Ernest picking up pieces of waterlogged sheetrock that had become so heavy and unable to support their own weight, they had ripped from the ceiling and crashed onto her sofa. Once that portion of the ceiling covering fell away, water poured freely into the room and sluiced over the backs of the men as they bent to pick up the broken pieces.

At that moment, the children screamed in unison, and Fiona could hear their bodies colliding with the door as they scrambled to exit the room.

"Wha' now?" Fiona fretted as she ran to the bedroom. The sight that awaited her was horrifying. Her heart stopped beating and the blood in her veins turned to ice. Where her three-month-old had been sleeping peacefully on the bed, even through the noise of the children playing, there was now an indiscernible heap covered in sheetrock. Fiona looked up into the darkness through the gaping hole of the roof and then back down at her hidden baby. The reality of what she saw before her could not penetrate her reasoning.

"Auntie Fiona, Evan is under there," her eldest nephew called from the doorway, stating the words that she was not willing to say herself, nor did she want them said aloud.

She reached under the mess, grabbed hold of one plump little leg, and pulled her child free, revealing his ghostly face covered in white paste.

Her breath caught in her throat; she feared the worst. He was not moving; she could not tell if he was breathing. He looked so lifeless, so still. She turned to face the children who stood blocking the doorway.

"Call Dan!" she croaked, but no one moved. "CALL DAN!" Every one of them scurried out of the room in fear that she would breathe fire like the monsters in their electronic games. She listened as they reached the living room and in chorus shouted the name of her husband.

Fiona turned back to her baby and prayed that he was okay. She did not know what to do. Her books on what to expect did not cover this. She was afraid to move him, afraid to touch him. *Dan will know what to do; he always knows what to do. Where is he?*

"DAN!!!" she shrieked in absolute blood-curdling terror. "It's the baby!"

Dan stumbled into the bedroom at full tilt in response to Fiona's call of alarm. She stood bent over the pile of rubble, her small frame shuddering as she gasped for breath. His eyes roamed to the bed as he tried to decipher the images of white mud, bed linens and other lumps and mounds that did not make sense. Suddenly, a movement caught his eye and Fiona's air exploded from her lungs in a burst. Little Evan kicked his legs and flailed his arms as he sucked in air and released a howl that threatened to shatter the windows. Dan rushed over to the bed and picked him up, hugging him closely to his chest with the masculine protection that only a father could exude.

Fiona stood listless, tears of relief running down her cheeks. She reached to brush off sheetrock that stuck to Evan's legs and back as Dan rocked in an effort to calm him.

"Oh, thank God, he's okay," Fiona sobbed, looking at the doorway full of curious children and concerned adults. Her mother pushed past everyone and joined her daughter in cleaning the baby. She reached up to take him from his father, who hesitated a moment before letting go.

"Let's run a bath for him and see if he has any cuts that need attention," she said calmly. Fiona was so thankful that she was there. She followed her from the room towards the bathroom.

Dan remained by the bed, alone, as the others dispersed to other areas of the house. He began to pick up the sheetrock, but was overcome with the shock of what just happened, or very nearly happened. He sat on the edge of the bed and buried his face in his hands. Quiet sobs emitted from him. He was not prone to crying, but this was turning into a very stressful day and he was just not prepared for any of it. The images of his son played across the movie screen of his mind, fueling the anguish within and resulting in more tears, which he suppressed so as not to draw unwanted attention. Dan took several deep breaths, roughly wiped the tears from his eyes and cheeks with the sleeve of his shirt, and got to his feet, gathering his full height. He would not let himself break down again. He had to be strong. Fiona needed him to be tough. Most of all, his precious son needed him for protection. He would not let them down, he could not.

SIX

Now the fear was there but whispering. Barely audible, yet still making its presence known.

"The electricity should stay on a little longer. The power company said that they would be forced to cut it only when the winds picked up to hurricane strength," Jackie announced, more to open conversation than to state a fact that everyone already knew. She had invited Michelle and Des to stay with her that same day. It only seemed natural to have friends over during something like this. Besides, Jackie's son, Naldo, was a good friend to Dylan and Demitri, and she hoped they would keep him entertained so that she could focus more on worrying. Jackie's sister, Patty, also lived with her and was not fazed by this sort of thing. She was very confident that she could handle just about anything and would be the one to deal head-on with any crisis that might arise. The atmosphere was tense with expectation.

"You just make sure that we have enough ice to pack everything on, once the power goes out." Patty called from her bedroom, "I don't want all that food spoilin'." She emerged from the room with a towel in hand, drying the rain off of her arms and then wiping up the droplets that had abandoned their human ships for the shiny tiles of the living room floor.

"Patty, I hope you get some rest tonight, girl. If this storm hits us with all it's got, you will be cleaning house for a month," Des joked, knowing that if the storm really hit with as much force as Michelle claimed it packed, there was a very strong chance that there would be no house at all come this time tomorrow.

For now, however, they were spared and everything remained operational. As the children were playing, oblivious to the true sense of the threat that was only hours away from reaching its full murderous potential, the adults busied themselves with checking doors and windows, then settled down to watch the television for as long as the powers that be empowered them to.

The mindless babble of the late night talking heads did little to suppress the growing intensity of their anxiety. Weather stations were intermittently discussing the threat to U.S. coasts and casually mentioning its impending effects on the Caribbean islands, which lay in its more immediate path.

"The Cayman Islands are next to feel the brunt of Hurricane Ivan. We continue to see people taking no chances here. They know this is a powerful and dangerous storm that has already killed more than two dozen people," the weatherman said solemnly. The words were so cold and factual. Des prayed that it would not be so fatal here.

By midnight, there was little else to do but retreat to their assigned bedrooms to rest for the day ahead.

Moving at only 9mph, Ivan was taking his time. Surveying the islands with a hunger, cracking his knuckles to the roar of thunder, eyes sparking with bolts of lightening. He was a predator stalking a wary prey that, while aware of his presence, would never be able to outrun his timed attack even if he lost the element of surprise.

The water company had announced that water service would be discontinued prior to the expected hit. Most everyone who heard about it immediately called family and friends to advise them to catch water in whatever containers they could. Bathtubs were bleached clean and filled with fresh water. Those who had cisterns ensured that they were full as well.

Fiona had taken the necessary precautions and her tub was almost overflowing. Her hands were shaking as she tested the temperature of the water that she dipped from the bathtub and poured into the infant's tub. She caught a glimpse of herself in the mirror as she stood in front of the vanity and was taken aback by the reflection that looked back at her. The eyes of that person were wide and haunted. She recognized the features, but the expression was so distorted that she had to do a double glance to be sure that it was herself she was looking at. Her mother held little Evan, waiting on Fiona to assure her that the water was all right to set him into.

Fiona stepped back, allowing her mother to move closer to the tub, which she had placed on the large vanity. Two porcelain sinks were under mounted to either side of long countertop, leaving a wide gap of unused surface area. She had placed baskets of potpourri and candles there to decorate the room and emit delicious blends of vanilla and cranberry that pervaded the olfactory senses in an array of scents. These were now lined against the back wall with her perfumes and some of Dan's colognes. The baby tub fit perfectly into the space and was just off the edge, so it could not tip over.

Too traumatized to help, Fiona stood listlessly beside her mother, who was working to remove the jumper Evan wore. Drying sheetrock mud caked the parts of his body that had been exposed, while the thin clothing had minimally succeeded in protecting his chest and abdomen. The removal of his outfit revealed pasty white mud that had seeped through the cotton, but it was mostly contained on the outer side of the fabric and the substance washed away easily as water was drizzled over him. His grandmother had to gently swab his arms and legs to remove the encrusted layers that had hardened.

The tears ran hot down Fiona's face as she stood there, watching her baby splash in the water. She gulped back the sobs that threatened to erupt from her. A hard knot caught in her throat and her mouth went dry as it wedged in her esophagus, like a dislodged boulder rolling down a narrowing crevice, cutting off the water supply to her mouth. Her graceful hands flew up to her face, trying to push the cry back inside. It was forcing its way past the boulder, shouldering around her tongue and banging on the inside of her lips. The sob welled up within her, powerfully insistent; it would not be imprisoned within her for much longer. She turned her body away from her child and faced the bathtub. Her eyes could not clearly decipher the shapes and colors before her as the tears continued to mount her eyelids like tidal waves, before running over and spilling down her face. They blurred her vision, but she cared not what her shower looked like at this moment. She was losing control and focused only on keeping her wail buried, which required no physical sight at all.

The battle was lost as she sucked air though her mouth and the howl found voice as it escaped. Behind her, Evan jumped and her mother grabbed her chest, both startled by the sudden explosion. Fiona's shoulders bent forward and her head dipped into her chest as she relented and succumbed to the emotional release. She swayed and her knees buckled, allowing her to crash to the floor in a ball of weeping mess. Her mother quickly grabbed the towel and swaddled Evan before kneeling beside her daughter and hugging her with a free hand.

They sat there for several moments, rocking; Fiona's mother hushing her with soft words of encouragement while simultaneously comforting the baby, who had joined his mother in a bawling duet. It was all Fiona could do to regain her composure.

The knock on the bathroom door was quiet at first, almost indiscernible. Fiona struggled to her feet, wiping the last smudges from her mascara-marked eyes. Black streams highlighted the telltale travels of her wayward tears.

Her mother had resumed the bath and was now drying Evan off. The knock came again, most persistent and louder.

"Who it is?" Fiona called, quickly dipping the washcloth in Evan's bath and then scrubbing a stubborn black mark on her cheek.

"It's Dan. Is everything all right in there?" the muffled voice of her husband called back through the locked door.

"Mummy, unlock dat door and let him in please," she begged her mother, "and go ahead and get da baby dressed." Her mother collected the clothes to be worn and dropped the soiled ones into the laundry basket. She walked to the door and turned the safety latch, allowing Dan to enter the bathroom as she exited.

"What happened?" he asked as she turned to face him. All mascara had been successfully washed away; however, she had not been able to fully conceal her bout of crying. Her swollen eyelids barely shuttered shades of crimson that were the tiny veins in her eyes. Dan reached out and took her left hand in his right one. He pulled her closer as she gently pulled back, tears instantly springing back to her burning eyes.

"Oh baby, don't cry," he soothed. "It will be all right." She did not resist again as he tugged her arm harder, but rather allowed him to draw her in to his chest and wrap his arms protectively around her petite shoulders. "We will get through this, do you hear me?"

She nodded weakly, the lump having returned to that crevice that was her throat. The more he tried to assure her that all would be well with them, the harder she fought to keep from breaking down again. Finally, she was able to take deep breaths and force the cry back into its cage within her chest. Dan took the cloth that she had used earlier and wiped the wetness from her face. He smiled at her confidently and she managed a slight grin back at him. He smoothed her hair and kissed her forehead. Sometimes he could be such a father to her.

"Take your time in here. The kids are all okay. We moved them to the other bedroom for now."

Fiona nodded without looking up. She sighed as she looked at the mirror to see what damage this last spell had done to her face.

Dan left, quietly closing the door behind him. She dabbed her cheeks a few more times and then flushed her face with cold water, letting her eyes soak, in the hopes that the red veins would shrink and the whiteness return. After drying with a hand towel, she gave herself a cursory once-over and drew in a final lungful of air before switching off the light and rejoining the family.

SEVEN

The cell phone rang deep into the night. The voice on the other end sounded masculine and very worried.

"Hello? Is this Des?" the person asked, panicked.

"Yes it is," Des murmured sleepily. The call had stirred him from the beginnings of his sleep.

"Do you know Elton?" the voice asked.

"Yes I do."

"I am calling from Cincinnati, Ohio. There's an elderly woman that is all alone and is very scared. Elton said that he would go there to be with her for me, but he cannot ride his bike all the way. He said that if I called, you would be able to carry him"

"Okay, I will see what I can do." Des ended the call and moved to get out of bed. Elton was an old friend of his who went by the nickname "Old Man." While he was happy to help him, he was very nervous about venturing out now. But he was more worried that it would rest heavily on his conscience if something happened to this old lady, and she had no one there to help. He thought of his own grandmother and knew that he could not bear knowing that she would have to endure a storm of this size alone. He decided to throw caution to the wind and brave the storm. The gusts were consistently raging now, and he could hear it howling outside.

He told Michelle of his intentions and quietly left the room, careful not to wake his sleeping princes.

He met Old Man a short drive down the road and noticed that the wind had already begun to blow trees down, littering his path. They drove in silence for a while, maneuvering around fallen branches and straining to see what was happening at the end of the beam of light that the headlights provided. The moon had been devoured by Ivan, a cheesy appetizer in delicious anticipation of the three-course meal to come.

The tall branches were waving excitedly all around them. A verdant flourish, rousing a chorus of windy applause and celebration as the two knights rushed to the service of the damsel in distress. Or could it have been a warning? The fervent gesticulation may have been merely an attempt to turn them back, to deter them from proceeding further. Either way, they seemed not to notice as they trekked on

"Man, this one is gonna be rough, Old Man," Des commented as he pulled into the woman's yard.

"Yeah, Dessie, I really didn't wanna leave mi yard, but boy, I kinda felt bad saying no. I hope that we make it through the night, you know what I mean?" He had a worried look on his face and a tremor in his voice.

Des knew exactly what he meant. "Well, g'night. I'll come by tomorrow and we'll take a drive to see what everything looks like."

Des drove away, the wipers on his windshield playing a furious game of tug-of-war. He passed house after house displaying boarded windows and shutters. Usually, this time of night played host to various walks of life along these streets. Tonight, however, no one could be seen. Everyone was tucked neatly into their boxes, praying that the wrapping would be secure enough to stave off the greedy fingers of Ivan, trying, like a disobedient child, to get a peek of his present on the night before Christmas.

He slowed as he neared the property where his horses were kept. Last count, there were nine. Des had always loved horses, getting his first one from his father on his sixth birthday. They were his passion, the very distraction that kept him out of jail when many youths were stealing cars and messing with drugs. His fears as a young man were that, should he ever do anything so stupid that it landed him a prison sentence, no one would be able to care for his horses and they would surely perish.

As he pulled into the side drive, the lights from the truck caught jewels flashing in the trees. The glinting gems bobbled up and down as they came closer to the vehicle like miniature UFO's seeking safe landing. Gradually, the headlights revealed the heads of several mares whose eye sockets encased the luminous treasures. They were all turned into the pasture to ride the storm out at their own risk. Des was very nervous for them, as they had never been through anything like this before. He wondered how they would handle it.

The matriarch of the group was the closest to the

truck, and she nodded her head up and down as if sensing his concern and trying to assure him that she would take care of rest. The property was very large and would afford the horses plenty of room to run if fleeing became necessary. As prey animals, equines were accustomed to taking flight when they felt threatened. They were quick on their feet, and this knowledge helped assuage Des's fears. He slowly reversed the truck back onto the main road and carefully made his way back to his family.

Des arrived back at Jackie's and breathed a prayer of thanks for the safe return. He kissed the warm heads of his sons and then returned to Michelle's embrace before succumbing to the blissful retreat of sleep, words of supplication on his lips.

Mama DeDe's tenant, Massoud, chose to stay downstairs in his apartment while she entertained her other guests upstairs. She had put her mother to bed earlier in the evening and then returned to enjoy the company of her friends. They had chatted for hours about their youth and the many adventures that they had been on. Even little Chloe was wide awake, the excitement too much to leave behind for the boredom of slumber.

With the aluminum awnings pulled down tightly over the windows, there was nothing to be seen through the dark glass panes. The house was well insulated, so the sounds of the ravaging winds were faint and whispered. Candles dotted each room, providing the haunting glow of conspiracy and intrigue. The flames licked at the air, eager for the taste of oxygen needed to sustain their fiery existence. As the electricity had been cut to the island, the adults had made light of the situation for Chloe's sake. They pulled out the supply of scented candles, tea lights and slow burners in all shapes, sizes and colors.

The shimmying flames danced on their wicks, illuminating each room and casting animated shadows around the walls and ceilings. All seemed peaceful and calm on the inside as Ivan wreaked havoc on the other side of the walls that shielded them.

Shortly after midnight, there was a knock on the inside door that led to the garage.

"Miss DeDe, I think you better come down here and see this," Massoud called out. His voice bore a tone of intense fear that she had never heard from him in the ten months that he had been renting from her. "There is a whole lot of water coming in the garage."

Chris and Al were the first to bolt from their chairs and catapult down the back stairs. Chris's arm nearly took out a figurine of a rooster that was perched precariously on the half-wall between the kitchen and the stairs. The rooster was frozen with a crow stuck in his throat for all eternity, but as it wobbled on its ceramic feet, it appeared almost to come alive and belch the "cock-a-doodle-doo" from its craw.

Mama DeDe's kitchen was a shrine to roosters. They were striding confidently across the curtains over her sink and staring inquisitively from the shiny surfaces of the decorative glossy ceramic plates that hung on her walls. Her fascination with them led her to seek them out wherever she shopped and made gift giving easy for her friends and family.

She even had the salt and pepper shakers, tea towels and place mats.

Now, every fowl seemed to turn in the direction of the stairs as the human contents of the house poured through the narrow opening to the garage below.

"Dear Jesus save us," Lacy murmured under her breath.

Just as Massoud had said, water was gushing under the door into the two-car garage in an urgent bid to saturate everything in its path. Mama DeDe pointed her flashlight around the perimeter of the room and watched in horror as small items that were not secured floated around, bumping into the larger ones that lined the walls. She had not parked her vehicle in here for some time, as it was more useful as a storage room, though it could comfortably house two cars, or in her case, one large Avalanche and any other sized car. Hers sat just beyond the garage door in her driveway, no doubt submerging inch by inch as the tides rose.

The winds had been driving everything down the street for hours - from lawn chairs to uprooted trees. Christy, a young mother of two, looked through the gap in the window, peering into the darkness and intently watching the waters rise each time the lightning flashed. The milky rains beat every tree into a shredded mess of leafless branches, which were pruned of their weaknesses by the unrelenting winds that were less gusty now and more like turbine powered expulsions. The sight was only visible to her when the heavens were illuminated by the sparks of electricity that frequently shot through the low clouds, casting a ghostly burst of light on the terrain.

Christy had long since given up on the hope of sleep. Her nerves were raw, her senses heightened. She barely connected her eyelids in a blink; afraid that the millisecond that it took to close and open her eye would be the very moment that everything around her fell apart.

Outside, the flooding from such a downpour was inevitable. With no outlet or drainage points, the ground was forced to pool the torrents where they fell, accumulating thousands of gallons most unnaturally.

She had focused her stare on the area where her small car sat in the driveway. Her vision did not shift even when the pure blackness of the lightless night swallowed up all sense of shape and form. Only when the skies lit up was she able to determine how high the water had risen. Her little car was now submerged to the window level in a sea of orange-brown, murky liquid. Her eldest, Ryan, a precocious youngster of four had earlier commented that, "…it looks like chocolate milk and orange juice, Mommy!"

Christy saw the humor in his innocent words, but could barely summon a smile to the corner of her twitching lips. She was in full-blown panic. As much as she wanted him to be asleep, snug and safe in bed with his favorite stuffed toys surrounding him, she was relieved that he was awake now and keeping her company. His bedtime had come and gone at least four hours ago, and she was sure that he would crash soon. Instead, he chattered to her about everything that came into his head; unknowingly keeping her grounded and alert.

In her arms nestled her newborn baby girl, barely two months old. She napped now, oblivious to the tension in her mother's grip around her tiny, frail body. Christy would have put her back into her bassinet as soon as she was fed, but at this moment she just wanted to feel the warmth of her daughter's head against her skin. While it would not have been unusual for her to be home with her babies on a Saturday night, everything bucked against the norm tonight; both in nature and in her home.

The baby's father did not live here. He had another home; with another family where he was surely doing everything he could to protect them. But not her. No, she was alone in her duty as the sole guardian of these two young lives; the only protection that they would ever know. She herself was barely an adult. What could she possibly know about protection?

The father of her eldest had at least come by with some plywood and boarded up her windows. Of course, it wasn't the best job, but then again, she couldn't have done it half as good. They were not on good terms, but it was thoughtful of him to see to her this way.

Candles flickered from various points around the small apartment. The glow transformed her modest place, rendering it cozy and inviting. She had been forced to light them around the time that she would have tucked Ryan into bed. The electricity went out as announced by the local service provider. This was a premeditated measure to ensure everyone's safety. It was at that point that she made the conscious decision to let her son stay up. The heat the candles generated did nothing to warm the chill that had settled within her. Her jaws tightened and her teeth chattered as her nerves responded to the heightened anxiety that she felt.

EIGHT

As Mama DeDe and her guests peered into the garage from the lower steps, surveying the room, frozen where they stood, the water rose. Nay-Nay suddenly swiveled on the step and began heading back to higher and drier ground.

"Chloe! Get back upstairs to Nay-Nay," Lacy shouted in panic.

"But Mommy…"

"NOW!" Lacy screamed. Her heart was racing at the vision before her. Maternal instinct was in overdrive and she was now a mother intent on preserving the life of her youngest child.

Chloe could be heard running up the stairs, each pound of her foot on the tiled plywood steps reverberating in everyone's chest to the rhythm of their own heartbeats. The mirth and jesting of the early evening had been replaced with fear and panic. Lacy was not sure what to do at this point. She wondered what was happening outside. The sounds down here were so much louder than upstairs. The thin garage door did little to hold back the howling and snapping that menaced just beyond its plastic slats. The cries of the winds and cracking of limbs certainly brought a sobering realization of how serious this situation was.

Lacy recoiled from the water's wet touch on her bare feet, moving backwards and upwards a step or two as Mama DeDe's flashlight spilled its beam closer towards them, losing its effectiveness in the swirling lagoon of dark and murky water. Lacy wanted out of this right now.

The children had gone to bed without much resistance. They were traumatized by the chaos that had ensued when Ivan's ferocity had somehow penetrated the inner sanctum of Fiona's home. Julia had fallen asleep with them in a pile of arms and legs, on the bed in the bedroom that had not suffered damage from any major leaks. Of the four bedrooms in the house, two were completely unusable. Dan had closed off the doors to dissuade anyone from entering. The sheetrock still lay on the bed in the room where his son, Evan had napped earlier in the evening. He had attempted to clean it up, but was so overcome with emotion that he was unable to complete the task. Buckets had been placed on the bed to catch the water that poured from the ceiling but little could be done to save the bed at this point. The closed door hid the sight and pushed it out of their minds temporarily.

Dan's mother, his mother-in-law, and Fiona's grandmother slept in the baby's room, which was the only other bedroom that remained intact, on his king-sized mattress. It had been retrieved before the leaks could destroy it and now lay on the floor in his son's room. Fiona was curled up with little Evan on the bottom of a set of twin bunk beds that sat in the corner of his room, perpendicular to his crib. The room was large for a nursery and easily accommodated everyone's sleeping arrangements. The bunks had been Dan's idea, even though Evan was much too young to sleep on them.

Fiona was not happy about having them in there with the crib, but tonight she had thanked him for forcing the issue.

Dan and Fiona's brothers did their best to manage the drips and gushes that perforated the ceiling and fell to the floor in each room. The three men took turns sweeping water, mopping water and dumping water. Their hands and feet were prunes from the excessive moisture, and their backs hurt from bending, pulling and lifting.

The house was in silence except for the thumping and pinging of water into the hollowness of empty buckets and pots. The men were sitting at the kitchen table, listening to the hurricane and its many voices, when the power went out, casting the house into deep darkness. Fiona had left the lanterns and flashlights on the table, which was centrally situated between the kitchen and living room. Dan felt in the impenetrable blackness in search of something that would bring light back to the room.

"It's a good thing the kids are sleeping," he said, depressing the rubber button that housed the mechanism to power the lantern. "Can you imagine the screaming and crying they would have done when the lights went out?" He chuckled, causing Ray and

Ernest to respond in kind. No one felt like laughing, but the thought of a house of screaming kids was much funnier than the thoughts of the house crumbling around them.

"Dan, why don't you go catch some sleep, and Ernest and I will stay up to keep an eye on things," Ray said, when the laughs had awkwardly faded to silence again. "We can take turns. We'll call you in a little while."

"Naw man, that's okay," Dan demurred. "I still got something left…for a little while longer, at least."

"Look, there's an empty bunk above Fiona, you take the first sleeping shift and we'll wake you up in an hour. That is, unless something happens sooner," Ray insisted.

Dan looked at both men for a moment then, realizing he would not win this argument, got up from the table and headed to the sofa.

"I don't trust that top bunk to hold my weight!" he said as he fell into the cushions. "I'll be fine right here and you won't have to risk waking anyone up to get to me." He fluffed a large decorative pillow with fringe and placed it at the end of the chair, then pulled his legs onto the end and slid down until his head rested on it. He wiggled and shifted until he found the right position, and was asleep in a matter of seconds.

The sandbags were holding back the bulk of the water, but the garage door was not built to provide a watertight seal. Mama DeDe could not even imagine what would happen had she not taken the time to make the bags and fill them with sand. She quickly hurried to the farthest wall where her belongings were stacked against the wall in boxes. Water splashed as her feet slapped the wet concrete, sending droplets in every direction and up the legs of her slacks. She reached for containers that rested on the floor, trying to put them higher on the pile. The bottoms dripped more liquid onto her clothes as she hastily grabbed box after box, trying to amalgamate the smaller stacks into one large one. Nothing could be done for the larger items that were fuller and heavier.

"Miss DeDe, I think you should go back upstairs now. I do not think it is safe for you down here," Massoud said, finally breaking the silence. "I will stay down here and make sure that everything is okay."

"I will too," Al stuttered. "Chris, you go back upstairs and stay with the women. We will call you if we need help."

Lacy and Chris began to climb the stairs, careful not to slip. Mama DeDe slowly turned to follow, but hesitated, her eyes meeting those of Al. No words were exchanged, but she knew Al understood what she was trying to say. She wanted him to stay safe and it worried her that he could not do that down here. She held her hand out towards him, flashlight extended for him to take. He nodded reassuringly, accepted the torch, and gently pushed her out of the way so that he could close the door between them. She retreated to the kitchen, heart racing, and shivering from the prolonged contact with the flood.

Nay-Nay met her at the top of the stairs, towels in hand, and helped her to pat her legs dry.

"DeDe, better you change out of these old wet clothes and put something dry on," Nay-Nay suggested.

"Yeh, I think you're right," she agreed. "You just sit here in the kitchen so you can hear if those boys call out for anything." She slung the damp towel around her shoulders, grabbed a battery-powered lantern, and headed towards her bedroom.

The hallway was the only area that did not have its own light source, but was sufficiently lit by teasing flames dancing on candles in each bedroom along its length. Hers was the last room at the end of the corridor, and had been designed with long, deep closets flanking either side of the entryway. She pressed the button on the lantern and the fluorescent bulb kindled within the glass casing, its bluish-white light expelling the shadows from around her. She quickly pulled a clean pair of trousers from a hanger and closed the door.

Each end of the corridor was anchored by a bathroom, one served the two forward guest rooms and one was for her private use. She retreated into the latter and closed the door behind her, resting the lantern on the top of the toilet tank. As she removed the soggy trousers that clung to her legs, she thought about her grandsons and wondered what was happening to them at this very moment. They lived in the western district, right across the road from the once-friendly sea in a single story home. Though it had stood the test of time for almost 30 years, she now questioned its structural integrity in light of the new developments of the storm.

She forced out the thoughts that flooded her mind and caused her heart to miss beats.

"They are just fine," she told herself, wanting to believe every word and imagining them sleeping soundly in their beds, oblivious to the madness that swirled around outside. She knew her son would protect them at all costs, and trusted his instincts to know what to do to survive.

Now dry and more relaxed than she had been in about ten minutes, Mama DeDe returned to the kitchen and turned off the lantern to preserve the battery. Lacy and Chloe sat in one chair, mother cradling daughter, while Nay-Nay and Chris each occupied their own. All eyes were turned towards her, wide and frightened. Gone were the smiles and jokes. The reality had pervaded their naiveté, extinguishing the playfulness and light-hearted antics as suddenly as a splash of water to a flame.

She slowly slid into the last remaining chair and placed her hands on the table, clasped in front of her to hide the shakiness that would belie her true feelings. No one said anything as each focused all of their senses on the hurricane outside.

NINE

Despite her attempts to remain awake, Christy had nodded off to sleep beside her son and daughter some time after midnight. They shared the one bedroom in her apartment, which was neatly arranged though somewhat crowded.

Rebecca stirred beside her mother and grunted as her empty belly signaled that it was time for the next feeding. Christy awoke, somewhat groggy and disoriented. She did not remember taking them to bed, and was concerned that she had not placed the baby in her bassinet beside the bed. It was not a habit that she wanted to practice. Ryan had no choice but to share the full-sized bed with her, but he was in no danger of being suffocated should she roll over too close to him. Christy sat up in the bed and pulled the baby into her arms.

The candles on her bureau issued low light as the flames struggled to hold onto the last bits of wick at the bottom of the jar. The liquid wax was all but burned away. She arose from the bed and reached for a spare candle, which she placed atop the dying one, quashing the flame.

"Shoot!" she mumbled angrily, as the baby's soft

mewls escalated into louder cries. "Why didn't I just light this one with that one?" Her mind was just too numb with drowsiness to think straight, she thought. She quickly lit the new one from the box of matches that she had purchased with the candles.

"Mommy," Ryan called softly, "whatcha doing?" She heard him shifting on the bed as she blew out the match and turned to him.

"Go back to sleep," she urged, Rebecca's cries drowning out her words.

"I'm hungry, Mommy," he whined, pitching off the bed and joining her. She put Rebecca high on her shoulder and reached for Ryan's hand.

"Well, come on then, baby, let's find some breakfast," she said, having no idea what time of the day or night it was. Together they walked into the dimly lit kitchen where the candles were suffering the same fate as those in the bedroom. She chided herself for being so careless yet again, this time in leaving the candles lit overnight. Not only did she need to preserve them for as long as possible, but she could have burned the house down while they slept. The thought made her heart race and cast the sleep far from her eyes once more.

Christy prepared the baby's bottle and placed her on the sofa, surrounding her with pillows so that she could attend to Ryan and rejuvenate the lighting with another round of candles. The refrigerator still held its coldness inside, even though the current had been turned off hours ago. She made sure to get in and out quickly, so as not to let the refrigeration escape. She poured cold milk into the bowl of cereal that Ryan had chosen, and served it to him with his favorite Mickey Mouse spoon.

With the room brighter, she returned to the sofa to be close to Rebecca, who had not even opened her eyes as she suckled noisily on the bottle. It was still dark outside and still stormy. She looked at her watch, which informed her that it was almost five in the morning.

She was sure that this would be an early start to a long day.

Hours had passed since the waters had begun to infiltrate the garage. Al was not sure how bad this was about to get, but he knew that he would sacrifice his life to protect those women upstairs, especially DeDe, who had been like a sister to him all of these years.

Ivan's abuse had continued through the night and was still full strength as morning broke. The storminess cast a meager dimness to the new day. Daylight barely filtered through the exposed panes at the bottom of the French doors that led to the garden behind the house. An awning had been pulled down to shelter as much as possible but it only managed to cover the top half of both doors. The remainder was disappearing behind the rising tide, leaving the room largely enveloped in the familiar darkness.

Al turned to Massoud. "Well, friend, me 'ope you 'av your bathing suit on, it look like we be goin' swimmin' soon-soon."

"I am afraid that I cannot swim," Massoud whispered, "Do you know if she has anything that I can float on?"

Al searched for a floating device that belonged to Mama DeDe's grandkids. "Put dis on," he said, "and whatever you do, don't h'open any of dem doors."

Massoud did as he was told and donned the swimming ring that Al handed to him. The two men stood in the middle of the garage, Al surveying the room corner to corner with the flashlight, and Massoud held tightly to his float, as if waiting for an impending wave that would sweep him away.

The water was rising faster now, the slats of the garage door barely holding anything back. It heaved from the pressure on the outer side, but still held its ground within the track to which it was fixed. Al trained the light at the top of the door so that the brightest concentration of light pinpointed into the fold between the ceiling and the wall, and away from what he wanted to see. The residual yellow glow illuminated the entire quarter of the room against the farthest wall, with lessening degrees of intensity marked in bands of light that etched outwards from the whitest center to fuzziest darkness. Cascades of dirty waterfalls could be seen cresting the garage door where the adjustable seams had compressed together when the door was put down, closing the gap between the inside and out.

The height of the spills gave away the true position of the depths of the waters outside, which was exactly what Al had wanted to determine.

"Friend, you see dat?" He nudged Massoud, pointing to the streams pouring in. "Dat deh a-tell you 'ow 'igh deh wattah a-reach 'pon the h'outside.

Effa we did stand 'ere in t'ree foot of wattah, me know say h'out d'ere mussa 'ave more wattah still!" Al's thick Jamaican patois accent was even heavier now. Massoud simply looked at him, not fully understanding the words, but the meaning was not lost on him at all.

"Deh h'ocean mussa meet inna deh miggle of deh h'island, man!" Al's hand dropped from its position, causing the beam to point downwards into the darkness of the water. Massoud's eyes followed, straining to see the bottom, but the light only glared off of the glassy surface of the rising pool, shimmering as fresh surges disturbed its upper skins.

The sudden creaking sound to their left signaled the opening of the inner door that led to the kitchen, and Chris's face appeared at the top of the doorframe. He stood on the middle of the staircase, bending and stretching forward to open the door in an effort to remain dry. The water quickly filled the stairwell and several steps, stopping just short of where he stood. The look on his face was a mixture of shock and fear.

"Everyt'ing okay?" he asked, though it was quite obvious that it was not.

"We're fine so far," Al replied. "How deh ladies dem?"

"They're doing okay," Chris said, glancing backwards to see if anyone had joined him at the top of the stairs. "You t'ink we're going to die?" he whispered.

Al let out a nervous chuckle. "Nah man!" he shot back quickly. "We in a two story 'ouse. Deh wattah could never reach dat 'igh!" He looked at Massoud, who quickly nodded in agreement, though his eyes darted this way and that, uneasily watching the water swirl higher around him.

Chris cautiously took a step down and then another, his feet enveloped by the rising tide. He waited momentarily, as he registered the change from dryness to wetness before continuing downwards. When he had reached the bottom landing, the water met him at his waist. He tugged on the door, pulling it against the water as he tried to close it. It resisted slightly, the motion of movement much slower than the effort Chris exerted would normally have produced. Al stepped forward and grabbed onto the side of the door, his strong, thick fingers wrapped around the edge as he forcefully yanked it towards him. It responded with resistance, but eventually they were able to shut it, sealing off any chance of any further breach to the upstairs.

TEN

Outside, the day had started afresh, but the Ivan's recycled winds and rain continued to wreak havoc, throwing trees, boats and random materials around in frenzied upheaval. Gloomy light struggled to penetrate the overhead blanket of thick black cloud, but not even the strength of the sun could permeate Ivan's armor. The ghost town that lay beyond Christy's window was desolate and wet. Inside her house, she continued to burn candles to scare away the dark shadows that settled in each room. The plywood covered most of the living room window, but was simply not big enough to fit across all of it. The large gap on one side afforded her a limited view of the chaos outside, each time lightning stretched across the sky. The slice of coverless window did little to light up the room, and the sunless sky with its dark, suffocating pillows of clouds ensured that the room remained cast in a nighttime eeriness.

The desire for sleep was completely gone from Christy and Ryan, but little Rebecca had no trouble finding it behind her eyelids. Christy envied the baby's ability to sleep through anything. She was such a good girl, not a fussy baby at all. She ate and slept mostly, and when she was awake she was so calm and easy-going. Her mother felt so blessed to have her.

Beside her, Ryan played with his toys on the sofa. Christy sat and watched him, admiring his imagination and maturity, while yearning to be that innocent and carefree again. Within the last half hour or so, the waters on the outside had found entry into the room through the badly sealed threshold of the front door. Christy only noticed this when the shimmer of candlelight had suddenly multiplied as the reflection of the flamed wicks stirred beneath her feet, catching her eye, and causing her to glance downward. The water on the floor in the living room of her small one bedroom apartment was pooling around the legs of the furniture, but had still not reached into the kitchen, which was directly opposite where she sat. The floor had been badly poured and there were many dips and bumps throughout the place. The linoleum covering was cracked and yellowed from years of abuse. Some parts had been ripped away, leaving cold bare concrete. It was cheap housing and, without consistent support from either father of her children, it was all she could afford.

The apartment was sparsely furnished. Most of the furniture had come with the place. It was cheap and old, but it served its purpose for now. Christy held down a full-time job at a local branch of an international retail bank and worked part-time for a fast food restaurant. The hours were long and grueling, but they still did not afford her the luxury of whim spending. She had no intention of wasting what precious little cash she had after bills and living expenses on more possessions that would only clutter this tiny place.

"Ooh, look Mommy, a boat!" her son, Ryan, squealed joyfully as he pointed out a dislodged vessel tossing around outside. He knelt beside her, legs tucked beneath him, rolling his small toy cars along the sill of the window. "Mommy, can we go on the boat?"

"No, baby, not today," she said soothingly, stroking his hair with her free hand. "Hey, why don't we go play with your cars in the bedroom after I put your little sister down?" She really wanted him higher above the rising tide, and figured that the bed was as good a place as any.

"Okay, but this time I want to drive the big truck. Vroom! Vroom!" Ryan mimicked, as he raced towards the bedroom. She loved her children more than anything in this world. They were the light of her life and the reason she could even get out of bed in the morning. Neither had been planned, and yet she could never have orchestrated such perfect additions to her existence.

Christy had never done well in school; was certainly not the one voted most likely to succeed. She was by no means the most valuable member of her team at work; in fact, she was certain that if she stopped showing up, no one would really notice. She lived an average life, just under the radar. Her reality was barely a ripple in the societal pool; yet she had given life to the most precious beings she had ever met.

Her son's eyes melted her heart when he turned his gaze in her direction. Never mind that he had inherited them from his father, the latter had certainly never caused her to feel such joy and pride. Little Ryan had stolen her heart from the day he was born. No man would ever claim it for himself again; and it certainly could never break for anyone now that its owner would belong to her forever.

The way he sought her out for comfort; asked her thoughts and opinion on every issue that perturbed him and loved her unconditionally — even when she looked like hell warmed over – every time he called to her simply reinforced that there was one person on this earth who *needed* her. She might not have achieved greatness thus far, but she was adored by someone so special. She was certain that she could never love another more than she loved her little boy.

Then along came Rebecca.

She watched as Ryan raced away from her towards the cramped bedroom to retrieve his toy truck. Suddenly, his bare feet slipped in the water and he went sliding across the floor, his little body colliding with the only bar stool and seat in the kitchen. He instantly erupted in screams of pain that startled the baby, provoking her to join in with wails of her own.

The young mother felt an overwhelming urge to follow suit. She was on the brink of losing her mind from fear. This was not her first hurricane, but it was the first one she had ridden out alone with two children. She felt completely unprepared and useless. She had sat at work, listening to the chatter of her coworkers as they schooled each other on what essentials to buy in order to survive a hurricane. With her limited funds, she had carefully evaluated each one, firstly in terms of necessity and then on cost. She had made the decision to purchase candles instead of a flashlight or lantern; generic brands of canned food over the popular names, and was resigned to the fact that her windows would be unprotected and exposed, had it not been for her son's father and his plywood. Her office had given out a few cans of tuna and batteries.

Those things were now of no consequence as she battled the hysteria that rose within her. Her attempts at preparation were paltry and lame. What on earth would a can of tuna do to save her from this?

She reached for the bottle on the side table and quickly pacified the infant, while simultaneously propelling herself upwards and outwards with her elbows towards her bruised son. Lightning blazed, and from the corner of her eye, Christy caught a glimpse of the ghost-piloted boat engine as it swung in the currents outside. As she watched in dismay, it lodged itself against her poor car, pocking its body with several dents. Her breath caught in her airway and her heart skipped a beat. She felt a deep sense of loss. That car was the only thing in this world that she owned debt-free. Her insurance would never afford her a new one worth driving.

Hearing her son's cries escalating, Christy moved to lift him up, but something slowed her progress. She felt that she was not walking as quickly as the energy she was exerting should have allowed her. As she looked down, she gasped in horror. The water that only seconds ago was skimming across the floor, was now mid-calf. Everything around her slowed to a mind-numbing fraction of real time. She became acutely aware of every detail surrounding her as she stood in the center of her living room and swiveled to look at the front door. The small imitation planter was no longer to the left of the door under the light switch, as it had been moments ago. It was now lazily floating past her on its side; the fake plastic leaves lapping in the choco-orange water as it gushed in.

The water had crested the lip of the window and was spilling into the apartment through the opening left by a broken pane at the bottom of the window, behind the ratty sofa. The light brown cloth darkened as the liquid soaked into it and quickly spread across the surface in a widening pattern, like a blossom opening to the sun.

Christy's eyes slowly moved beyond the sofa and window to gaze at the wayward boat that still docked by her car. So much water was pouring in, yet the level outside was not receding. In fact it appeared that the tide was rising as the water fought gravity to inch upwards against her windows.

The flood that continued its course through the broken pane now filled the gap entirely, as if a large funnel was plugged into it, siphoning the putrid juice from the outside into her home. The glint of the candlelight on the coffee table next to the window cast a honey glow onto the exposed glass. A wall of nasty brown liquid swirled past the casement, pressing against it with significant force. Leaves and bits of debris floating on and in the river, never stopping long enough to rest against the panes.

She continued in the dream state for an eternity, unable to do anything until she was jolted by screams so shrill that she just knew someone had to be dead. She turned back to her son, who had climbed up on to the barstool. Ryan's mouth was closed but he looked at her with wide-eyed shock. The sounds were not coming from him, they emanated from within her. She had no idea how long she had been standing there screaming, but the water had risen to her thighs, extinguishing the flames of the candles behind her and darkening the face of her frightened son.

ELEVEN

Mama DeDe and the other ladies sat in the kitchen upstairs in the early morning hours, bantering light chatter as the winds prevailed outside. Each had taken turns at sleep, but deep slumber was impossible for them. Mama DeDe's mind's eye surveyed every corner of her home, even as she engaged in the chitchat. Her thoughts were everywhere, but she was especially concerned for the well being of her grandchildren.

Everyone paused occasionally to listen intently to the sounds of the wind soughing around the house. The radio sat on the countertop, the constant drone of the announcer passing on important information. It was battery-powered, and Mama DeDe had more than enough batteries to keep it going for weeks.

Trees could be heard slamming against the outer walls or snapping altogether. Every once in a while, something would hit the roof with freighter force, but no one could determine what it had been.

Al was downstairs with the other men, but Mama DeDe knew that he would come back up if she needed him to. As she thought of him, she heard footsteps on the stairs and saw his head appear around the corner, a look of worry on his face.

"Miss DeDe, de water fe rise fas' down deh," he said, his patois thick and heavy. "Unna better stay up yah, me nah want you fe get sweep 'way."

"How high is it?" Mama DeDe asked, rising to her feet and moving towards the stairs on which he stood. Her truck, her washing machine, all of her tools and everything else one would put into a garage were down there. It was beyond her reach and at the mercy of the hurricane.

"Boy, me would say it mus' be 'bout four anna 'alf feet 'igh," Al responded, lifting his hand to register a mark of measurement against his chest. Mama DeDe did not realize it before, but now she noticed that his clothing was soaked.

"Me goin' go back to dem downstairs, but you stay 'ere." With that Al turned to descend, not giving them a chance to argue. Mama DeDe cautiously placed one foot on the next step before jerking it back quickly and returning to the kitchen table.

The ladies sat in silence for a while, processing the information. The gales outside preached a mighty sermon. More debris rocketed against the house to drive home a point, not of fire and brimstone, but tempest and downpour.

The gunshot that next ripped through the house brought a deafening roar. The women shrieked and ran to the stairs, all warning of the flooding forgotten. Mama DeDe stood frozen, her eyes fixated on the hallway that led to the bedrooms. She grabbed the flashlight and trained it in the direction of the sound. The kitchen sat on one side of the foyer, which led into the hallway. From her vantage point, she could clearly see the front bedroom situated at the opposite end of the house. Her light revealed the ceiling buckling with the force of the wind.

Lacy bent over the half wall that separated the kitchen from the back stairs and yelled for the men.

"We have to get everyone in the back closet," Mama DeDe shouted. Though she had no formal military training, she suddenly knew how to command and at this moment, each of these ladies needed someone to tell her what to do.

"Lacy, help me get Mama up and let's move her first," she said with authority. "Nay-Nay, take Chloe and the lantern and go into my big closet in the hallway."

Beatty had been put to bed in the middle bedroom and no doubt heard the explosive crack from the room right next to her. Indeed, when they reached her she was sitting upright in the bed, looking expectantly towards the doorway.

Mama DeDe rushed to the bedside and gently but firmly coaxed her to put her legs over the edge so that she could help her mother to her feet. Lacy joined her, reaching forward to grip under the elderly woman's arm and back as she lifted.

The ceiling in this room rippled and flapped as the winds chewed away at the wooden roofing above them. Mama DeDe dared not look up for fear that acknowledging it would only speed up the destruction and she would then be sucked out and tossed aside like a ragdoll.

She and Lacy heaved and pulled, drawing Beatty

out of the bed and momentarily holding her until she steadied herself. The older woman looked into the face of her youngest child with eyes that clearly did not register the significance of the danger she was in. The lines of wrinkles, produced by years of worry and fret, and to a minor extent by aging, were permanently etched on her face but not enhanced by the current situation.

For almost a year now, she had only maintained a look of serenity and peace, her mind slowly receding into itself as Alzheimer's took her away from her family one brain cell at a time. Her ability to understand basic instructions diminished daily.

Now Mama DeDe was trying to tell her to walk forward, shouting above the increasing volume of the engine that churned overhead. The old woman took a step forward and stopped, again searching her daughter's face with a blank expression.

Wood splintered and rain pelted as the edge of the roof was torn away above them. Lacy screamed and ducked as something catapulted toward her. Suddenly Al appeared and muscled between Lacy and Beatty, grabbing the elderly woman and lifting her into the air in one deft movement. He cradled her and bolted from the room, ladies in tow, just as the sheetrock collapsed into the room and onto the bed.

Mama DeDe closed the door behind them and stumbled forward towards the closet where her sister waited, phone to her ear.

"Put Mama down in the corner over there," she instructed Al, as she surveyed the size of the closet in relation to how many bodies could fit into it. Clothes and boxes filled the room, leaving little space for anything else. Al set Beatty down and stepped backwards into the hallway to allow Mama DeDe to enter the closet. Nay-Nay and Chloe had already nestled themselves against the back wall, and Lacy was shifting things around so that she could move closer to her young daughter.

Mama DeDe looked back to the doorway, but Al was no longer there. She worried that he and the other men would not find shelter downstairs, but knew that he would do what he had to in order to survive.

Unable to sleep due to the roar of the wind that had replaced the soft whispers earlier in the night, Des, Michelle and the rest of the household slowly began to come to life as each member arose and gravitated toward the front windows, where they could get a look at the action outside. Preparations on this house involved nailing sheets of plywood to the outer side of the windows. Some sheets were smaller than the windows and therefore allowed the view of the front yard on either side of the frame. The battery-powered clock showed that it was just after four a.m., therefore, whatever view could be had of nature's version of the Olympics was obscured by a thick black blanket of darkness. No moonlight to cast a ghostly glow, no streetlights to illuminate the world beyond. All power had been cut and the explosive exhales of Ivan the Terrible had snuffed out the internal candle of the moon. Pitch blackness, so dense that even a cat could not adjust to it, penetrated through the layers of the earth. The sounds were the only evidence of life on the planet. The wailing and howling of the wind, the cracking of the limbs as they snapped under the weight of the driven rain and wind, bowed down to the ground in a humble reverence offered to appease the wrathful ire of the king. Ivan was climbing in ranks, ascending to his throne in a swirl of chaos and turmoil. The trees and plants kow-towed to his outbursts, trembling like servants under chastisement, fearing death.

The reports stated that Hurricane Ivan would pass to the south of Grand Cayman – the largest of the three islands, barely sweeping its murderous eye within 35 miles of the southern coast. While this was a welcomed alternative to actually receiving the full blow of the eye, seasoned islanders knew that the damage would be only marginally lessened, though no one would ever have imagined how much a partial Ivan could vitiate these islands.

As the hours wore on, the occupants of the house that had become a prison alternated between chatter, watchful vigil, and eating. The light of day, though greatly diminished by the intense cloud cover, revealed a scene that was both frightening and awesome. The rain no longer sheeted lazily down as it had done the night before. Instead it now pummeled horizontally, as if the stories of the old world explorers had come true; the earth was actually flat and had turned on its side, causing the sky to settle in the east. The oceans were raining their contents onto the land.

Wind and rain combined to create snow, or so it seemed, as the sight that lay beyond the portal of vision was hazy white as in a snowstorm. Houses across the lane that were once fully visible were now enveloped in the milk of the hurricane. Even the trees in the front yard were barely evident, their brilliant colors washed away by the freighter of wind and rain. They stood as if behind an opaque shower curtain, outlined in gray.

The landscape was dotted in black. Portions of material lay in the open grass lot across the road, barely identifiable through the rain. As piece after piece was added to the ground, creating a mosaic of odd shapes, it became apparent that the shingles were flying off of the roof above them and being flung to the earth. They became projectile objects, slicing through the air and pitting the ground. Watching them in amazement, Des and the others began to realize that this was not a good sign.

"I pray that this passes over quickly. There is nothing more than tar paper and plywood underneath those tiles, and if the wind could loosen nailed down shingles, it would not take much more to chew through the rest of the roof," Des surmised as he paced back and forth in the living room.

The mood began to shift toward fear as everyone contemplated his words. What had begun as a day of boredom, isolation-wrought tedium, was now being transformed into minute by minute of anxious uncertainty. Nervousness registered on each face with its telltale signs of urgent upward glances at every noise.

Suddenly a phone rang, breaking the verbal silence that had been thick in the room. It was Des's cell phone and he answered it guardedly, without looking at the screen to see whose name was listed on the caller ID.

"Hello?" he said hesitantly. "Hello?"

He moved around the room, trying to get better reception as the women followed him with their eyes. Worry hung on his face like a wet rag, dragging his features down in gravitational stress.

"Aunt Nay-Nay, is that you?" Des stopped in the middle of the living room and tilted his head to the right and then to the left. He found a clear spot and froze in mid-movement like a marionette that had been abandoned by its puppeteer.

"Slow down, I can barely understand you. Where's Mummy?" He listened to the voice on the other end of the connection with rapt attention. His eyes widened as the information that poured through the line into his ear found its way to the part of the brain that converts sound to sense.

"Where is everyone now?" Des's lips pursed to blood-draining whiteness. Black circles appeared under his eyes with such swiftness that to blink would have missed the transformation. "Stay in there, don't cry. Stay in there. Tell Mummy that I love her…yes, I love you too…no, you are not going to die." He wiped the sweat from his forehead and straightened his back, which had begun to cramp in his awkward position, losing his reception in the process. He quickly dialed his aunt's number, but the circuits were busy and his call would not go through. He tried his mother's number, but to no avail, everyone in Cayman must have been trying to call their loved ones to say goodbye.

TWELVE

The front bedroom, which no longer offered protection from Ivan, amplified the howling and whistling of the wind. The sounds penetrated the closet that housed the four adult women and one child. So tight were the quarters that the closet door refused to close entirely as their bodies pressed against it. Mama DeDe fought with it, trying to pull her mother in closer to the wall, but to no avail. She sat nearest the opening and held on to the stubborn door with all of her might.

The heavy breath of Ivan was invading the hallway that led from the damaged rooms to the closet. Its salty halitosis assaulted their nostrils. The house shook and trembled as it was now being attacked from the inside as well as from outside. The wooden walls threatened to rip from the floor and join Ivan in the air like a perverse spin on the biblical rapture.

They crouched in the closet for an eternity, or so it seemed, as each second was measured by a burst of air though the house, unrelenting and constant. Mama DeDe heard a voice from within the deeper regions of the closet and strained to understand over the roar of the wind. She recognized it as Nay-Nay's but could not make out the words so clearly.

"Dessie...the roof...closet...your mother...gonna die...love you..." The words that she did hear were stuttered and broken but Mama DeDe got the point. The call had gone out to her son. She was glad to know he could answer the phone. That meant he was all right, and more importantly her grandsons were too.

Suddenly an angel appeared in the doorway, a deep mocha colored hand extended to Mama DeDe. It was Al. He had returned and was resisting the pummeling of the wind to help them to safety.

"C'mon Miss DeDe, you can't stay in deh," he shouted, pulling her to her feet and then reaching around her to grab Lacy's extended hand. Nay-Nay grabbed up Chloe and turned to look down at her mother.

"S'ok," Al reassured her. "You go on, I got her."

Al bent down and pulled Beatty to her feet, and then once again lifted her into his strong arms as if she weighed no more than a pillow. With everyone out of the closet, each held on to one another as they made their way to the stairs. Mama DeDe glanced into the room where the sound of a freight train was roaring through. The roofing had been completely lifted and tossed aside by the giant ogre that was terrorizing the neighborhood. Above her head, the ceiling sighed and moaned as the wind passed through the opening between the rafters, trying to pry the remainder from the house.

Mama DeDe knew that eventually all of the rain would leach across to other rooms. She knew that the preservation of her personal belongings now relied entirely on how well she had packed them.

As the group walked the remaining few feet to the stairs, she looked back, trying to memorize the way it looked, because she knew that by tomorrow this scene would be drastically altered.

Al had burst through the downstairs door that led from the garage into the house when he dashed upstairs to check on the ladies. It remained open now, forced to the wall by the water that had risen beyond the eighth step. They slowly walked down each step, careful to place their feet firmly to avoid slipping. Al supported Beatty in his arms, as her ailing legs could never have carried her with the surety that everyone else was afforded. One by one they stepped into the muddy water, its cold wetness pressing into their pores.

Mama DeDe waded into the garage, the water reaching her at her chest as she searched for Massoud and Chris. She found them by the washer and dryer, which were now entirely submerged underwater. Massoud was atop the washer, clutching the floating device.

Chris hugged the wall as he came towards them, offering assistance to Chloe, who was too short to walk through the water and keep her head above it at the same time. He pulled her to the machines and placed her next to Massoud.

Mama DeDe's garage was large and spacious, built to hold two vehicles. Two individual doors were electronically controlled to open at the front of the house, designed to roll upward into a storage casing at the top of the garage. At the other side, and directly opposite these doors, hung a pair of French doors, which opened to the back yard. It was through these two sets of doors that the swells were breaching the security of the house and invading the downstairs.

For a long time now, Mama DeDe had preferred to use the garage for extra storage, so it was filled with boxes and containers from ceiling to floor. There was still plenty of space to move around, as she always tried to stack things against the walls so that it never looked too cluttered.

Now, however, all of her painstaking organization seemed to have been in vain. All around them items floated and dipped, rearranged by the surge that filtered through the openings of the garage.

Al batted away an errant box that drifted into his path as he clutched the elderly woman in his arms. The height of the water buoyed her body, making it lighter to carry her than force her to walk. Chris approached with an inflatable air mattress and held it steady while Al hoisted Beatty onto the makeshift raft.

And so they waited, Nay-Nay and Lacy poised by the washer and dryer that were tucked into an alcove in the wall, Mama DeDe, Chris and Al steadied the floating airbed on which Beatty reclined, oblivious to the surrounding turmoil. The threat of losing the roof on the house was much greater than the water rising up around them, and yet with every inch that it crept, the end drew nearer.

The situation was quickly becoming unbearable as the reality of what was coming pressed into the minds of each adult. Chloe sat beside Massoud, panic and fear evident in her cries. Each adult fully understood her panic and envied her youth, which allowed her to express her emotions without a second thought.

Just when it seemed that there was nothing left to think about but the all-encompassing water or the few moments of life that they had left, it began to slow and withdraw. With a suddenness that took everyone by surprise, the water began to recede back through the doors. No more than ten minutes had passed since they first set foot into the garage and now it was being sucked back out.

"Thank God!" Nay-Nay cried, a sense of relief settling on her face. Each person repeated the two words, their worries sloughed away with the lessening tide. Mama DeDe fought the urge to fling open the French doors, which now acted as a dam holding the water back. She wanted to throw them wide and hasten the drainage, but knew better. Outside the winds and rain were still very much present and very deadly. She watched as the floating items gathered at the exit points, pressing against the doors, unable to squeeze through cracks and gaps.

Though the water was a large body when accumulated in one place, and though it could kill a person with no effort other than its persistent embrace, its very make-up allowed it to separate from itself without damaging the whole. It could remove portions in a drip and be rejoined later, not altering one aspect of its original form. It was capable of passing through the tiniest of holes, where the naked eye could not even discern an opening, and pool back into the same matter on the other side. This body of water that filled the garage was quickly extricating itself through miniscule gaps and slots and teaming up once again with the river that passed through the streets outside.

Like a liquid Passover, this death-seeker traveled house-to-house, looking for more than just the first-born. It sought the unprepared. No victims were here, however, so it moved on, following the tug of its master. Ivan pulled the watery chain and the obedient torrent responded, abandoning this home for another and then another.

"It's over!" Mama DeDe rejoiced, "We made it through alive!" She reached down to pull her mother up. Beatty still sat atop the air mattress, which had slowly descended to the floor with the reducing tide. Al and Chris helped her heave the elderly woman to her feet as Mama DeDe straightened out her wet dress. Beatty's hands joined in, smoothing the fabric around her thighs, though it was hard to tell whether she consciously did it or was just mimicking an act.

"Don't get too over-confident just yet, DeDe," Lacy cautioned. "You hear that outside, we still have a while to go before this is over with, and at the rate that your roof is going, we'd better stay alert."

"That's true, girl, but what you see over our heads is pure concrete, so I am not worried in the least about it falling down on us. We are in the best place as long as the water stays away," DeDe replied, drawing all eyes to the ceiling above them. The entire downstairs portion of the house was solid cement, unlike the top half, which was constructed of Texture 1-11 wood and insulation. Apart from the vulnerable doors, the garage was a bunker that proved far safer than any other part of the house.

"Jump on the counter, baby!" Christy called to her son as she regained her composure, the water threatening to steal her balance and send her and the infant splashing into its muddy clutch.

"But Mommy, you said you would spank me if I ever did that again," Ryan cried between sniffles. His eyes held hers in a fearful stare, not quite sure if he should be more scared of the water or of her punishment. He rubbed his bottom and right side, where he had connected with the floor and chair when his foot slid in the water.

"I know baby, but it's okay today, just this once." She laboriously waded over to him and hoisted herself and her newborn onto the countertop, careful not to knock over the arrangement of candles that she had strategically placed there earlier the night before. Her jeans were heavy with the nasty liquid that had invaded her humble home. They clung to her legs like shrink-wrap, sealing off everything from her calves downward. The extra material folded around her ankles, gushing water into and over her soaked tennis shoes. The shoes were no longer brilliant white, as the canvas absorbed the rancid water into its fabric, instantly dyeing them a dingy brown. As she drew her feet under her to sit cross-legged, the water squished out from her footwear, creating a puddle on the countertop and saturating her rear.

"Isn't this fun, honey?" Christy tried to laugh, but the sound of her voice was more frightening than funny. "Remember the game we play at the beach, 'Don't let the water getcha'? Well, how about we play that now?"

"I don't want to. I want you to make the water go away," he sobbed, his tiny voice trembling. "Mommy, make the water stop."

"I can't honey, we have to sit here and wait for it to stop raining." Christy had never felt so helpless in all her life. She could only try to diminish the true severity of the situation for him so that he would not become any more alarmed. At four years old, Ryan was a bright little boy. He was adept at sensing her moods and understanding what she was not saying. He was the "man" of the house, and as such he had taken on the parenting too, much to her surprise. He was so much older than his age at times, that when he would slip back into childish play, she had to put her disappointment in check. Christy had lost her childhood at sixteen when she conceived him. Above all, she wanted him to enjoy his.

The rising tide was now lapping at the underside of the counter's ledge. Soon they would have to stand to keep from being swept off. She surveyed the room to identify where the remaining candles still bore their flames. In the kitchen, one on the top of her refrigerator danced and flicked, high above the water, and four sat on the L-shaped countertop. The living room was dimmed by the loss of the ones that had rested on the coffee tables, but a single blaze persisted against the farthest wall on a small knick-knack shelf that was nailed to the wall beside the front door. Christy was relieved to note that it held one of the larger candles with three wicks, certain that it would last through the day.

The refrigerator suddenly tilted forward and crashed into the water, sending a tidal wave over the boy's head and almost knocking him over the edge. She instinctively stood up and grabbed his arm, jerking him to his feet. The canisters by the stove that housed sugar and flour were sent crashing over the edge of the counter by the surge and began to float away toward the back door that led from the kitchen to a very small, rocky backyard.

The little boy bent to pick up a container that held his favorite cereal, but his mother pulled upward so that his reach was shortened and he could only watch in sadness as it floated away from him.

"I'll buy you another box tomorrow," she comforted him, "and I promise you can eat as much as you like when I do."

His upturned face was almost angelic with gratitude.

"Can I get the one with the cookies and marshmallows instead?" Ryan pleaded.

"Don't push it, buddy," she said sternly, but with a smile, "You know that is way too much sugar for you."

He smiled and even let a giggle out. The sound was so pure and sweet that Christy thought for the first time everything would be all right. The eruption had removed the small candles from around their feet and atop the fridge. The margin of daylight that filtered through the small opening of the window and the three-wick candle cast the only light in the room now, not quite meeting the glow that teased out from ones in the bathroom and bedroom. As the water continued to rise, she knew that the latter would soon be snuffed out, too.

"What happened? Who was that on the phone?" Michelle asked as she finished pouring milk into a bowl of cereal that she was making for Dylan.

Des hesitated until his son took the breakfast and exited the room before speaking. "That was Aunt Nay-Nay. She said that the roof just ripped off and the garage has filled with water. They are all crammed into the back closet but couldn't get the door closed." He stopped speaking and walked over to the window to peer out into the world beyond, as if expecting the wind to blow the words that he searched for into his head. He turned back to Michelle as Jackie and Patty stepped closer to hear.

"They called to say goodbye, they don't think they'll make it through the day."

The women gasped collectively at the thought. Nay-Nay was Des's favorite aunt; he would do anything for her and she for him. Yet now, just when she needed him, he could not be there for her. His mother was there also and it was breaking his heart that she had stubbornly insisted on staying in her home when he plainly told her to come to his instead.

Mama DeDe would not have it. She would rather go down with the ship. Now his anguish grew to anger as he thought of his defenseless grandmother. She was 84 and barely able to walk anymore. He envisioned the struggle that they would have to pull her out of the two-story house and swim to safety, if that was what it came to. His heart beat with the hooves of a thousand horses pummeling his chest.

"Baby, as soon as this storm breaks, we'll go there, no matter what. They'll be fine, don't worry; we'll all be fine. Have faith in God...that He will protect us." Michelle hugged Des as she spoke softly, comforting him.

"Yeah, Des child, they're okay." Jackie tried her best to sound upbeat. "You know Nay-Nay, always overreacting. She's got drama!" She reached out and patted his arm.

Des managed a light smile, but his face did not relax, nor did the black circles fade from beneath his eyes. The atmosphere in the room was dark and apprehensive.

Within moments, this was overshadowed by the sound of dripping water. At first it was indiscernible through the roar of the winds outside, but as it persisted, it was unmistakable – gentle plopping sounds of liquid kissing a flat, hard surface. While largely known to trigger a very calming feeling over a person, thought to be therapeutic in its constant, timed rhythm, this new sound only succeeded in bringing a higher level of terror. The splashes were initially several seconds apart, but gradually became a symphony of sounds as one after another a new seepage opened up. There was a leak somewhere in the house, and if not dealt with soon, more problems would begin to arise.

THIRTEEN

September 12th, 2004
0400 hours
Ship call sign: ATPN
Hurricane Report: Latitude 16.8 Longitude 80.2
Approximately 190 miles SSE of Grand Cayman
Category: 5

Walls of water were driven skyward by the force of the winds as Hurricane Ivan made his way to the coasts of the Cayman Islands. He frolicked in the middle of the Caribbean Sea, a delighted child with newfound strengths. Ocean liners and other sea going vessels were warned long ago to steer clear of these waters at the risk of being capsized. The only unfortunates were the sea life that remained unaware of the dangers of the tempest.

As the hurricane circled around itself, its girth easily measured 400 miles across. He was a force of nature by all descriptions.

Wave upon wave rolled from every side in circular motion from his inward energy. Each swell peaked at eight to ten feet high before brute winds sheered them off and back into the sea to be collected by another wave. Ivan's whips extended as extremities from the outer limits of his body and sought to draft a new regime of water militia that continued to revolve around him like a swirling hoard of angry wasps armed only with stingers filled with salt water.

Fish struggled to dive to depths that freed them from the deadly grasp that would send them flying through the oxygen rich air above the safety of the waters. The inexorable commander and his hydro-minions misappropriated the unlucky ones as missiles and held them captive in the upward draft.

His full land strike was mere hours away, yet already he was wreaking havoc on the shores with his lashing tails. Drumming up waves and rain, he was teasing the residents with a taste of worse to come. Ivan enjoyed this cat and mouse game, felt supreme control as he instilled fear in his victims. Let them run, let them hide. He would find each and every one of them and by the end of this day, his name would be forever burned on their lips with a salty aftertaste.

Patty was the first to reach the hallway where the water was already pooling on the floor. The tiles were of high-shine quality and very slippery when wet. Des had joked the night before that someone could slip and fall in a teaspoon of water on these floors. At the time, the joke had evoked several laughs. However, at this moment, no one appeared to express any amusement.

"Jackie, get those kids in Naldo's room," Patty ordered with an air of authority. "Des, grab that mop in the garage, please, we need to keep the water under control or we will have some real problems on our hands."

Des agreed and hurried to the door at the end of the hall that led to the garage. As he pushed on the door, Jackie's dogs scrambled to charge the house, fear oozing from their eyes like energized molasses. The garage usually housed one car. However, for the time being, it was home to Max and Toby. The car, along with all other vehicles, was parked in the empty lot across the lane, equidistant from all trees and utility poles.

The garage door rattled like dry bones against its hinges. Ivan was ensuring that anything unsecured would be his loot from this invasion of the island. He tested the strength over and over, as if not believing that it was really tight enough and would be his prize with one more windblown shake.

Des quickly grabbed the collars of each dog before they could raid the house and kicked the door shut behind him.

"Easy boys," he soothed as he stroked their fur. "It's okay, you're safe in here." He thought of his own dogs, which he had enclosed on his porch behind concrete walls and aluminum shutters. He could only imagine what they were experiencing. He prayed that they were all right and vowed to run home the instant that the winds died down. Grabbing the mop, he negotiated his reentry into the house without the company of the canines, surveying the garage door through the small window in the house door to assure himself that it was not being carried away by the gusts. He returned to Patty, who was throwing towels on the floor with Michelle and Jackie.

"If you have some buckets, we can put them under the leaks," he said, as he looked up at the ceiling. The glistening pimples of water could be seen best by the flashlight that Michelle was pointing upwards. The ceiling had taken on the appearance of a diamond-encrusted covering above their heads. All along the center, running from end to end, little bulbs of liquid popped and released their contents to the floor below, some tagging body parts as they leapt to the ground to be reunited with their brothers and sisters.

"Jackie, didn't you add on this section of the apartment right about here?" Des questioned as he continued to look up worriedly.

"Yeah, Des. In fact, if I'm not mistaken, the original eaves of the roof came to the center of this hallway." She gasped with the sudden realization that slammed to the forefront of her mind.

When she had originally built the duplex, both sides were designed equally to comprise of one bedroom and one bathroom. Years later, she had extended the south side to include two additional bedrooms, another bathroom and a one-car garage. The present hallway was made up of the outer wall of the old structure joined by a roof to the new side. Now, there was an obvious flaw in the work that was done to make the two as one. It had not been seamless, as she had always thought.

Des was quick to sum up the situation. "What's happening up there is that the shingles are not holding back the water. Somehow, it's getting in and running down the old roof. The runoff is pouring into this spot and unless it stops raining soon, the sheetrock is going to get so heavy with water that it will fall – right into the hallway – and onto someone's head!"

No one wanted to hear this right now. The day had been so uneventful, so boring. Sure, they wanted excitement, who didn't, but not at this cost. Michelle opened the door to the boys' bedroom and peeked in on them. They played, oblivious to the seriousness of the conversation that was whispered just inches beyond their realm of fantasy. She longed to be that age again, just for this moment, so that she would not have to be privy to this information. Escapism was not part of her make-up, she could usually deal with adult issues, but at this moment the reality of being an adult was not as appealing as that of being a child. The prospect of having important decisions made for you and being able to play through a crisis without fear called to her. She quietly pulled the door shut and returned to mopping up the floor with the towels, which were already fully saturated.

Soon they would run out of dry towels and if Des' prediction were right, a simple bucket would be no match for the downpour that would threaten to wash them away from the inside.

The trauma of the previous night was still fresh in Fiona's mind as she awoke early on Sunday morning to her son's cries. She felt like a train had slammed every one of its iron cars into her diminutive frame, one after the other. Sheetrock lay all around the house. The house was abuzz with activity as the men had done their best to remove the larger pieces, leaving the women to mop up the water that poured steadily from the holes above. Fiona's and Dan's mothers were busily attacking the overflow of a bucket that could not be emptied quick enough to catch the constant drainage from the ceiling. The children played quietly in the front bedroom, where they had slept through the night. The excitement was long gone and they were nervously watching the ceiling to see if it too would fall. The storm continued to rage outside, winds howling by at break-neck speeds.

Fiona had woken with a skull-splitting headache and launched headlong into depression over the sad state of her house. She handed her baby to her grandmother and asked her to prepare a bottle and feed him.

Dan lay on the sofa, his arm swung over his face to cover his eyes, his legs crossed at the ankles. The raw emotion had taken its toll on him, and despite his words of reassurance to his wife that everything would be all right, he was now in an unresponsive, comatose state. He refused to move and did not answer when spoken to. This worried Fiona, who was still shaken from the ordeal of the night before. She sat on the edge of the sofa, where the curve of his waist provided more room for her petite bottom to fit.

She leaned in to him and whispered his name, but he did not answer. In the bleakness of the artificially lit room, the rise and fall of his chest betrayed him, as it was not the deep, timed rhythm of a sleeper. She knew this from spending countless hours watching her baby rest.

"Dan, please do not do dis to me right now," she pleaded quietly, "I cannot do dis alone." He did not move, nor did he respond. "Speak to me!" she urged in a stronger, more forceful tone. He remained still, refusing to give in to her demand.

Fiona let out a sound of disgust and pushed away from him, rising to her feet. She stomped off towards the kitchen to pour herself a drink and calm her nerves. Her hands shook as she gripped the cup and slowly poured the juice into it. Suddenly a head peeked around from behind her, startling her slightly. An impish grin peered through the tumbler that she filled, distorted by the bluish tint and wavy pattern on the exterior of the plastic.

"Auntie Fiona, can I have something to drink?" The twisted cerulean eyes fixed on the rising meniscus as she continued to pour, stopping the flow when it had reached about an inch below the brim. She handed the beverage to her nephew, whose features returned to normal shape and complexion now that she no longer viewed him through the blurring walls of the glass.

The little boy gulped noisily, juice spilling around the sides and down his chin. He finished the drink and let out a long sigh of satisfaction as he placed the glass in the sink. He turned to leave, but as he brushed past Fiona, a sharp cracking sound rang out from over their heads. Fiona jerked her head upwards in time to see the long rectangular light fixture ripping away from the ceiling and threaten to crash down on top of the two of them. Instinctively, she shoved her nephew aside and dove towards the living room.

The child yelled out and Fiona screamed as the sounds of shattering glass filled the house. Water that had filled the deep well of the light covering splashed everywhere, as the shards of debris flew in every direction.

"DAN! Get up! Help us! Oh Heavenly Father, please help us!" Fiona lay on her stomach, arms splayed in front of her, covered in water and broken glass. Her nephew had been thrown to the floor by the force of her push, but had rebounded quickly and shot out of the kitchen, even as the light was falling. He was at her side now, unsure of what to do next. His eyes were wide and frightened, tears were sliding down his face, but he was otherwise unhurt.

FOURTEEN

As they stood on the kitchen countertop, wet and shivering, Christy wondered if this was to be the last day that she and her children would live to see. She looked down at Rebecca, her infant daughter, nestled in her arms. She was still in a deep sleep that defied reason. How did infants sleep through such craziness? The baby was so tiny, so defenseless, and naturally relied on Christy to protect her.

Christy remembered the bittersweet day that Rebecca was born. That labor was much easier than the one she had endured with Ryan, yet the pains still caused her to double over and cry. The father was not there for the delivery. She assumed that he was enjoying a quiet evening at home with his real family. He had told her that he had car troubles and was not able to make it to the hospital when she called to tell him that her water had broken. Of course, that didn't explain why he never showed up to see his child until she was a week old.

She ceased caring about what he did or did not do months ago. He was not thrilled at the prospect of her pregnancy and urged her to get rid of it, but Christy refused. He was afraid that his wife would find out and leave him, but Christy did not care. They had a horrible fight and he walked out, only to return the next day with flowers and lame apologies. He wooed her until she gave in, and promised her that he would be there for her and his child. She believed him.

Now, as the waters surged through her tiny apartment, Christy berated herself on her ignorance and naiveté. She had put the lives of her children at risk by being so bull-headed, and now they would all pay.

Her son stood next to her, his arms wrapped tightly around her left leg. He looked up at her, searching her face for a sign of assurance that everything was going to be all right. Ryan knew that she could not lift him up in her arms as well, but he had still asked her to. Her heart broke as she had reluctantly told him no. He clung to her with all he had, his little body trembling with fear and frigidity as the waters rose to their ankles and then their calves.

Christy looked around at her home, watching the furniture rotate and float in the room. Everything was being driven toward the kitchen as the mild currents from the door and now from the windows caused objects to flow away from the source. Her television was completely submerged, as were the few pictures that were meagerly placed around the living room. She had not chosen to be miserly in her decoration; it was simply a matter of economics. Christy needed every penny she earned to sustain her children, and that left nothing to splurge with.

She watched as the water inched up the walls as if it were a creeping vine seeking sunshine. She felt so helpless and unable to move, unable to call for help. The tenants in the apartment next door had left the island at the first announcement of the hurricane. They had packed up and caught a plane to Miami without so much as goodbye. She could not do that. This was her country; first of all; and she would feel like a traitor if she abandoned it. Mainly, however, she simply could not afford it.

A young married couple occupied the apartment on the other side of her. They had left the day before to stay with the wife's family in the western district.

Christy was glad that everyone had somewhere to go, unlike herself. Her two children were the only real family that she had anymore.

This young mother had to do something quickly to preserve her babies' lives. She had to think quickly, come up with a plan to ensure that they would remain safe, even if she perished. As the thought occurred to her, she was already in motion, carefully sliding her feet along the surface of the countertop so as not to slip and fall. She transferred the baby to her right arm so that she could brace her son with her left. She reached the wall that housed the overhead cabinets and the stove, and shifted the infant back to her left arm. Christy selected the cabinet that contained the cups and glasses and began to remove them, tossing them over her shoulder and allowing them to land wherever they fell. Her son, still tethered to her leg by his own grip, watched in silence as she grabbed as many as she could with her free hand and threw them behind her. Ryan's head turned from side to side as he observed the volley of drinkware.

The shelves were deep enough for the purpose that she sought, and once cleared, she was able to rest her napping infant on the bottom row. Unlike the others, this one had a raised lip at the front that would provide a more stable cradle for its precious occupant. The base was covered with a flowery adhesive lining that she had picked up at the hardware store when she first moved in.

This was only temporary bedding, as it was still too close to the rising surge for her satisfaction. Rebecca was wrapped tightly in her blanket and therefore had enough cushion behind her head to ensure comfort. The cabinet was of the double-door variety, which had also factored into her selection as she was able to position the baby's body in such a way that her tiny head peeked out from one side while everything below her neck was concealed when the other door was closed. This offered another layer of protection to her helpless child. She tucked the baby's bottle into the folds of the wrapped blanket.

With the weight of the infant removed, Christy was able to extend her arm, which had gone numb. She stretched and cracked her back and neck by twisting her body to the left and then to the right, and rolling her head around in a circular motion. Satisfied that she had put everything back in place as best she could, she surveyed the topside of the cabinetry that faced the ceiling. Whoever had built the cabinets had been more concerned with depth than height, as they appeared to be abnormally deep but were only equipped with three small shelves. It was never a cause for concern before, as she did not own very many dishes anyway and therefore had quite a bit of unused space left.

Now, in the dim glow of the lone candle across the room, she eyed the top. She tried to evaluate the safety factor of perching her son there for a while. The distance between the top of the cabinet and the ceiling was enough to accommodate the little boy, and from this angle, she felt that it was wide enough to allow him to sit or lay, as he preferred. Her only concern at this point was the question of it holding Ryan's weight.

"No! Put me down! I don't wanna!" Ryan kicked and wriggled, trying to extricate himself from his mother's grip.

"Please, honey. I just want to put you up higher. It won't be for long, just until the water goes away." Christy had been struggling with him for about two minutes. His little hands pressed firmly against her neck and chest, trying to push away from her. She stole a glance at Rebecca, fearful that the commotion would awaken her, but the infant slept soundly, oblivious to the protests of her older brother.

She thought it would be easier to convince him to do something a little out of the ordinary. How often did he get to climb on the counters? Never. Ryan always responded to getting away with things his mother usually did not allow. Like the time he was allowed to eat ice cream before bed. That was a big no-no, but if Mommy said it was okay...

Christy couldn't tell him that it was because his father was a bad person or that she had made a mistake getting mixed up with him in the first place.

She couldn't admit that the ice cream was more for her than his little sweet-tooth; a quick fix for a bad mood, not a treat at all.

"Listen, I promise I won't leave you, okay?" She held him closely, cupping his head against her belly and lowering her head so that her lips were close to his ear. "I will stay right here with you the whole time. I just want you to get out of the water so you don't get sick.

"Remember when you caught that cold the other day because you wouldn't come inside from the rain?"

"Uh-huh," Ryan mumbled, tilting his head upwards to look at her. "I felt snuffly."

"Yes, you did, and Mommy had to make you take that yucky medicine, remember?"

"Yuck! It was nasty. Ewwww!" He used the back of his tiny hand to wipe across his mouth as if to remove a lingering droplet of the horrible medicine he had been forced to consume so many days ago.

"Right, so let's do what Mommy says so you don't get sick again, okay?" she coerced, as she simultaneously hoisted him up onto her hip. This time Ryan offered no resistance and instinctively wrapped his arms around her shoulders to hold on tight.

Christy took a deep breath and with every ounce of strength she could muster, she lifted and pushed him above her head, one hand under his arm and the other supporting his little bottom.

"Climb up now," she said, as she took a step closer to the cabinets. Her feet fought the resistance of the water that was once again circling her calves.

She looked down to examine the scene and felt her son's body teetering in her grasp as she lost focus and concentration. Quickly she turned his attention back to him and spread her feet farther apart to reinforce her stance. Ryan was looking down at her, fear fixed on his face, one hand clutching the top of the cabinet.

"It's okay, sweetie, I was just fixing my feet," she said, attempting to comfort him. "Okay, I'm going to lift you higher so you can throw your leg up there and pull yourself up. Can you do that?" Christy tried to keep her voice even-toned. The last thing she needed was to have him throw another fit. If she could just get him up there, then he would have no choice but to stay, because he would be too afraid to move.

At last he swung his leg upward, his foot catching the lip, and slid his body onto the topside of the cabinet. Ryan laid there on his belly, looking down at his mother, a small smile of relief and accomplishment at the corners of his mouth.

"Good boy! Brave boy!" she cooed. "You are such a big man now!"

With one hand bracing him atop the shelf, she slowly turned her body, careful to keep her feet solidly planted on the countertop beneath her. The murky water swirled around her knees; cold and wet. She had to collect her thoughts, had to come up with an escape plan. There was too much water, just too much water!

"What will I do? How can I get them out of here?" Her mind was so disoriented. She could not force her thoughts to cooperate and reason out a solution.

The level of the water on the inside was still lower than that which pressed against her window outside. She could no longer see the gaping hole of the broken pane. It had disappeared beneath the surge of water.

Staying here had been such a bad idea. Not that she had many options in the first place, but even a shelter would have proven more reasonable than this. How long had this gone on for, anyway? They had been enduring this torrent all night, and it showed no signs of letting up. Surely the hurricane had passed over by now. Christy had been through a few in her short lifetime, and it was all over in a matter of hours. Was this one just circling around overhead? Had it doubled back on itself in a perpetual circuit that simply could not be broken, like a sticking vinyl record?

Something tapped her on the top of her head, trying to get her attention. Looking up, and half expecting to see her son's hand reaching for her from his perch, she was slapped in the eye by a missile launched from her ceiling. She bent her head forward to wipe her eye of the wet residue and deflect another attack. Hand still clutching her face, she straightened back up and looked towards her son, who was intently staring at the spot over her head, just as more bombs found their target on her crown. Christy took one step away from the cabinets towards the edge of the countertop and raised her eyes to see what had caused the dripping. The ceiling was crying droplets of water from a large section of the sheetrock that covered the kitchen. Its underbelly swelled, straining the small nails that held it in place and buckling the now flimsy boards outwards. The seeping tears glistened in the dim light and suddenly gave way to a constant flow that met the awaiting pond that was once her home. The light fixture in the center of the kitchen was quickly filling with water as well.

It occurred to Christy that this was no longer a safety zone for her children, but before she could move to gather them up, the largest section of the ceiling released from one end of its mooring and crashed into her with a powerful punch. Christy was launched forward into the frigid waters, even as she heard her son's cry of warning.

Dan remained on the sofa, unmoving as his wife called out for his help. He heard the crash of glass in the direction of the kitchen, but he was unable to move his body or even turn his head to see what the commotion was about. He struggled with his emotions as they waged a vicious war inside him. He needed to get up and help his wife, at least see if she was all right, but a deep heaviness held him back. He was unable to move, his body felt like lead. He felt the inertia and knew that he must not let it overtake him, and yet it did. He tried to fight his way out of it, but his body rebelled against every thought to move, refusing to allow him to get up. His mind raced and yet his being was lifeless.

Fiona screamed for him again and then again. He felt the change in atmosphere as someone rushed past him to help her. The current shifted away from him, taking his breath with it. Fiona was yelling at him, but he could not make out her words. He was sinking into a dark place that did not register human sounds.

Suddenly, someone was throwing his arm away from his face, grabbing him by the shoulders and shaking him. The darkness that enveloped his vision was fading to brightness. His eyes began to concentrate on the murky shape in front of his face that was haloed by light. An angel. His angel. The features of his wife came into sharp focus and he smiled, just as a glass of cold water was thrown into his face.

FIFTEEN

The cold waters enveloped Christy as her body slammed into the rising tide. She did not have a chance to catch her breath when the falling sheetrock knocked her from the countertop, leaving her son and daughter unprotected and vulnerable.

Oh Lord, please don't let them be hurt! she thought as she sank to the floor, her eyes and mouth tightly closed to prevent the rancid water from getting in. She pulled her legs under her body and kicked upward. Her head broke the surface and she gasped for air. A sudden current was drawing her away from the kitchen, towards the front window. The water was withdrawing. As her body hit the wall, she was already fighting the current to get to her children. She tried walking against the flow but was simply pushed back towards the window. Pieces of furniture drifted up to her and she reached for a floating planter, which she used to smash out the remaining panes of the window to expedite the drainage. The shards of glass did not immediately fall to the ground as the plywood on the other side partially held them in place. The water, however, was free to gush out and escaped through every angle possible. She imagined that the badly installed threshold had finally loosened itself and washed out, leaving a significant gap beneath the door.

She turned to the kitchen again and was horrified to see the water pouring through the gaping maw of her ceiling. The cupboard was hidden behind a hanging piece of sheetrock, preventing her from seeing her children. The current quickly ceased as the level of water fell below the window's opening. The rest of the water would either slowly retreat through the gap under the door, or all at once when someone opened the door. Christy did not care at this point; her only concern was the safety of Ryan and Rebecca.

She frantically fought the resistance of the remaining water, desperate to reach her children. As she finally touched the counter, ready to jump up on it, a loud banging erupted from the front door. Christy spun around to see a man's face at the strip of window not covered by the wood. He was peering in, his head covered by a rain slicker, his bare hands cupped around his eyes to shield them from the bright glare of the flashlight that bounced off of her window as he shone it into her mostly darkened home.

"Miss, are you all right? We are here to rescue you," he called above the noise of the wind, which still persisted outside, though it had died down somewhat. His clothes whipped to and fro, pulling his body side to side as the force of nature called Ivan attempted to whip him from his feet with tendrils of air. He turned and said something to another person who was out of Christy's line of vision.

She could not speak, her voice caught in her throat, full emotion ravaging her mind. She needed to get up on the countertop and see what had become of her babies. She also needed to unlock that door so those men could get in here and take them all to safety. Christy struggled with the decision of what to do first. She could not bear to think of what lay beyond that hanging sheetrock. Not a sound came from behind it.

"Miss, please open the door, we are here to help," the man shouted, as his companion once again banged on the door. Christy looked from the man to the cupboards and back again. She slowly released the edge of the counter and wading through the lessening water to the front door and unlocked it. She bent to the window and with only inches between her face and that of her rescuer's, said, "I have to save my children! I cannot see them anymore."

As she hurriedly turned away, the second man jerked the door open, allowing the remaining water that had been trapped inside to pour out. The deluge almost knocked him from his feet. Likewise, Christy was put off balance by the sudden change in the standing water to another rushing exodus. She lurched forward to reach the kitchen before she could be dragged out of the door. Reaching up, she was able to grab and pull her body onto the counter once again. Two beams of light now darted around the room as both men entered with their flashlights and surveyed the living room and kitchen.

She quickly yanked on the broken ceiling, tearing

it from the rest and dropping it as soon as she felt it release. There lay her son, just as she left him, his face buried in his arms, his knees drawn up under him so that his bottom was elevated. Ryan turned his head to look at her as she placed her hand on his arm. His wide eyes spoke his fear with a voice louder than sound could relay. He noticed the bright lights that moved back and forth and look back at her expectantly, both hands extended for her embrace.

Beneath him, in her private quarters, Rebecca lay on her back, still tightly wrapped and deeply immersed in her infantile dreamland. Her twitching lips were evidence of her vitality. Christy breathed an exhausted sigh of relief and felt her legs begin to liquefy. A strong hand grabbed her as she sank downwards. It was her rescuer, the one who had peered into the window. The other man catapulted himself into the place that she was vacating and reached for Ryan who hesitated momentarily before accepting the hand and allowing himself to be removed from his safe spot. He was momentarily set on the countertop, but quickly removed to the floor by the other man.

Rebecca was lifted and carefully pulled down to floor level and placed in her mother's eagerly awaiting arms. Christy covered the baby's face with kisses and held her close to her chest. The infant stretched and yawned, her back arching and her face turning bright red. Her tiny face puckered as a cry of hunger issued from her open mouth.

Christy laughed as the sound of her infant rang out. Nothing sounded sweeter at this moment. The baby bottle was still wrapped in the folds of the blanket, so she was able to quickly retrieve and uncap it. Rebecca hungrily attached to the silicon nipple of the bottle. Although it was cold, she greedily ingested the milk with satisfaction. Christy wrapped her hand around the baby's body so that the same arm that held Rebecca could also hold the bottle in place. She bent down and pulled Ryan close to her chest, hugging him fiercely. He returned her hug, his small body quivering and shaking, his fists clutching her wet hair tightly.

"Miss, we need to get you to safety," her hooded rescuer urged. He gently pushed her towards the door, but she resisted, putting her hand on his arm to stop him.

"I need my handbag," she said, but he shook his head and resumed his push.

"No ma'am,' he urged, "You have to leave everything and come with us."

"It's right in the bedroom," Christy pleaded, "Everything I have left is in that bag! All the money I have left." Her eyes filled with water as she searched his face for a glimpse of understanding. "I cannot leave it here. I have nothing left."

The man released the pressure on her back and nodded.

"Be quick," he urged, but she had already slipped past him and made it to the hallway. He hoisted Ryan into the air and perched him on his hip, turning to face the hallway that Christy had disappeared into.

She immediately noted that the door to the bedroom was closed as she turned the corner into the hallway. Hers was a single bedroom apartment that just barely fit her full sized bed and dresser. Ryan slept in the bed with her and she was able to park a small bassinet between the closet and the bed for little Rebecca.

She shifted her daughter to her left arm and leaned into the door as she turned the knob with her right. The door was not locked but it held fast in the door jamb, unrelenting against her pressure. Christy used her hip to try and exert more force but the door would not budge. She turned back to the living room where both men stood waiting for her; Ryan reached out his arms towards her, fingers grabbing air as he beckoned her to return to him.

"I can't open the door," she said, raising a finger and nodding her head in Ryan's direction to signal that she would be there in one minute. "It's stuck."

"Miss, we really have to go," the window man urged, looking worriedly through the partially covered window, "We don't know if the flood will return or what could happen next."

"Hang on, dude," the other man interjected; then to Christy he said, "Let me try it for you." He was the one who had banged on her door. A man of little stature vertically, he was broad and solid. It was obvious he worked out on a daily basis. As he walked towards her, his arms swayed back and forth at 45 degrees from his body, unable to rest in a full perpendicular hand from his shoulders; his biceps bulged too much to allow it.

The doorman reached her bedroom and with one hand on the knob and the other pressed against the top portion of the door, heaved inward as he reeled back and then slammed his weight into the wooden portal. Without stopping to see what progress he had made, he repeated the act three more times in quick succession. The door gave slightly on his first attempt, water poured from the crack. On number two and three it opened wider, but offered great resistance from whatever pinned it in place on the other side. The fourth attempt succeeded in opening a gap that was big enough for Christy to squeeze through, however the new deluge of pent-up water prevented her from entering.

She and her rescuer stood against the wall, allowing the drainage to finish. The lack of the vacuum effect created by the receding waters beyond her front door meant that this water could not be carried away. Instead it pushed past their feet and sought out all crevices and recesses to invade. When it had finally tapered to barely a flow, Christy entered her bedroom with one of the flashlights; baby still nestled in the crook of her arm. Her eyes focused on the state of her private quarters.

"What did you do that for?" Dan sputtered as he sat straight up in the sofa. Cold water drenched his face and chest. His soaked shirt cleaved to his upper chest, which had taken the brunt of the deluge. The cloth that hung loosely around his waist remained dry only until the threads of running water detoured from their streaming paths and absorbed into the thirsty cotton. Wetness seeped across the folds, seeking to darken the material by several shades.

"What you was tryin' to pull?" Fiona demanded angrily, poking him in the chest with her index finger. Her other arm bent outwardly from her body and then back in as her wrist pressed into her hipbone, leaving her elbow jutting out sharply. Her stance was threatening; he had seen it before a time or two and knew that he was in trouble, but could not readily understand why. "Did you t'ink dat you could jes curl up and go to sleep and everyone would jes' leave you alone? The house is fallin' down, Dan!" She drew his name out for two beats, exaggerating the 'a' sound. "But I guess you didn't notice dat with your eyes closed!"

Dan wiped his face with the towel that Ray had handed him. He tried to remember what was happening and what would have caused her to go off on this rampage. He looked to Ray for help, but he just scratched his head and turned away, not wanting to get involved. Ernest was standing in the doorway to the hall, but when Dan shifted his eyes to him, he backed away and disappeared.

"Fi, I...I dunno what happened, baby."

"Don't 'baby' me, Dan. 'Baby' nothin'! I could've gotten killed jes' now. Where were you? Huh? Wha' were you doin'?" Her eyes blazed in fury. She was ready to fight. Fear had brought her here and if he did not come up with something quickly, she would do more damage to him that the hurricane that raged outside ever could.

"I came to lie down, that's all I remember. Really!" he explained defensively, sensing her unbelief. "Wait. I remember hearing you talking to me, but I couldn't answer." He locked eyes with her, wanting her to see that he was serious.

"C'mon Dan, you could've answered me. I was sitting right there."

"Honestly, I wanted to, but I just couldn't," he said insistently.

Julia was standing nearby and stepped closer to hear. She sat down in the armchair across from the sofa. "You know, I read about that somewhere. That if someone gets so scared about something, they can kinda blank out and go into a semi-coma. But it's like you can see their eyes are open and they look normal, only they don't really see or hear you."

"Yeah, that's what it was like at first," Dan agreed, eagerly hoping that Fiona understood, but more so that she believed. "I could hear but I just couldn't get out of it and answer. I am sorry." His eyes were as clouded as the shroud overhead, beyond their ailing roof, above them in the bruised heavens. Fiona's countenance gave way to a slight relaxation. Her jaw line, which had been taut with anger, slackened. Her eyes no longer sparked with the white-hot current of live electricity, but rather drooped as the heat of the moment subsided. The burden of the recent events of the day began to weigh heavily on her, and soon the fight was gone.

"Dan, just help us clean up the mess, please," she said tiredly. Fiona dropped her hand from its wedge against her hip and let the arm fall listlessly to her side. She walked away towards the kitchen with Julia in tow. Dan pressed the towel to his face once more and exhaled fiercely into the thick fabric. He wiped downwards, dabbing his chest and shirt.

Outside there was no quietness, only the continued surging of the wind in its inexhaustible quest to remove everything in its path. Dan walked to the front windows and peered through them. There was no view of his front yard beyond, only the dark underside of the metal shutter could be seen, pulling and pushing as the wind lifted and pressed. He imagined what lay beyond. The sounds were louder over here. The 'swooshing' was amplified with fewer concrete walls to diminish the noise. Something outside rolled around, crashing into solid objects. He could not determine the source of the collision but knew that it was large enough to inflict significant damage. He backed away from the window, as if a sudden premonition foretold of that very object crashing through the shutters and window.

In the kitchen, Fiona and Julia began to pick up the pieces of broken glass. Ray entered with a mop, while Ernest fetched the broom from the utility closet. On the floor inside, he noticed several bottles of bleach among other items one would expect to find in a broom closet.

"Wha's with all dey bleach?" he asked loudly, knowing that the vacuum-like void of the small room would muffle his words to unintelligible mumblings by the time it traveled to the nearest ears.

"I'm not sure, Ernest," Fiona answered back. "Mummy always stocked up on bleach during hurricane season, so I t'ought I would too."

"It fa affa dey storm," their grandmother answered

as she rounded the corner of the hallway that led from the bedrooms. "Fiona, dey baby is asleep again."

Ernest closed the closet door and joined the others, carrying the broom. "What good it is affa dey storm?" he asked, voicing the confusion that everyone else felt.

The elderly woman stooped down alongside her granddaughter and began to collect glass. Her hands were just as graceful as Fiona's, but not as long. They also differed from her granddaughter's in that they bore signs of years of hard labor. The fingernails were kept short and the skin on the underside was creased and rough. The tops of her hands were the shade of dark mocha, sans the latté and raised veins and bony relief spoke to agedness.

"Ma son, if God spares life, I hope you will not hafta find out," she said. "All's I knows is, when dey's no 'lectrici-dey and no runnin' wadder, it gets mighty hard to kill dey germs and all dat." Her accent bore the strong dialect of her roots, which had also been imbedded in the vernacular of her offspring. Originally from the eastern end of the island aptly named East End, she had now lived in the central capital for many years. But the native tongue of her district was still bold, as if she had never left.

"You needs it to clean out t'ings, to flush dey toilet, to kill the mildew. T'ings like dat," she explained as she continued to collect the larger pieces of the broken light fixture and place them into the garbage bag Ray offered. Dan had edged in on the group and stood listening with interest as his wife's grandmother gave her educational tutorial. He wondered if the bleach would become necessary to the degree that she meant.

SIXTEEN

Fiona looked up at her husband from her crouched position when she saw his feet step into view. He tilted his head slightly towards his left shoulder and then made a sharp nod, angling his chin toward the hallway as if it were a third arm, pointing away from him. He did not lose eye contact with her and she immediately understood his signal. She placed the shards in her hand carefully into the bag that Ray still held and rose to her feet, stretching her back, which had begun to adopt the hunched position, into a new posture. She delicately stepped around the last remaining pieces of kitchen light casing and bulb and followed his lead.

Dan stepped into the bathroom and waited for Fiona to cross over the threshold before closing the door behind her. He took both of her small hands in large manly ones and looked deeply into her eyes. Brown iris met brown iris as he searched for the right words to convey his feelings.

"I am so sorry, Fiona," he said with every ounce of sincerity he could muster. "I really cannot explain what happ…"

"Shh," she said gently, placing her fingertip

against his lips. "*I* am sorry Dan, for gettin' so angry. I shouldn't have talked to you dat way. My nerves are jes' so rattled." She hugged him, wishing that the storm outside would soon fizzle out so that they could get out of the house and escape the menace of cabin fever that was becoming a threat. He hugged back and kissed the top of her head. Though not voiced, they each knew that the other had granted forgiveness, and the comfort was like a delicious milkshake that slowly slid down into your stomach, coating everything with sweet happiness.

Fiona opened the door of the bathroom and stepped into the hall. The sounds from the bedroom where the children played were muted and conspiratorial, like an underground poker room where only the well informed knew the password for entry. She took a step closer to the door, which had not been properly closed and peeked in through the inch-wide gap. The children huddled in a circle on the floor. A variety of ages were represented, but the eldest held the gaze of each of the younger ones, who stared into his face with awe and admiration. The older child was telling about the destruction that this hurricane could cause, weaving a story that rivaled the works of the best authors in its perilous prose. His younger siblings were transfixed as his arms stretched wide to exaggerate the size of something and then slapped together to accentuate its impact. Each child jumped, the toddlers inching closer to the older ones, who put a protective arm around them. Mouths were open and eyes bugged wide as the children sat mesmerized, unable to leave and risk not hearing these pearls of information from this obviously knowledgeable scholar, yet petrified to learn more about the horrible fate that awaited them.

"All right, dat's enough!" Fiona reprimanded the oldest as she pushed the door all the way open. All children were startled by her sudden intrusion, shrieked, and clambered to their collective feet as if caught in the act of some taboo activity. They raced around the edge of the bed, bumping it repeatedly and causing little Evan to jostle as he lay sleeping in the center, surrounded by pillows. The movements only succeeded in rocking him deeper into his slumber as he stirred not. Fiona wondered why her grandmother had brought him into this room rather than the quiet nursery.

"You know better den dat," she chastised the storyteller. "Mind I don't call your mother in here to clap you!" The boy hung his head, embarrassed that his powers had been usurped in the presence of the other kids.

"Auntie Fiona, are we going to die?" A little hand tugged at her hem as the equally tiny voice of her four-year-old niece begged the question.

"No sweetie, we not gonna die," she said with more cheer in her voice than she thought she could feign. She stole a cutting glance at the eldest, narrowing her eyes to sear through him as if they were ocular weapons capable of firing lasers at will. "Don't you listen to him, he's a bad boy." Her words seemed to bring relief to the other children as she heard them each sigh as a group, glad that someone of higher learning than their previous infantile professor could dispel such gloomy tidings.

"I am going to fix somet'ing to eat for all of you. Come so wash your hands in dey bathroom." She took the hand of the four-year-old and led them all down the hall in the hopes that by maintaining the normalcy of basic routines, they would not be beckoned by the paranoia and hysteria that was reserved for the adults alone.

Outside, the wind picked up even more and the roof almost seemed to throb from the pressure. She hurried the kids into the small bathroom and pulled the door until it almost shut behind them, leaving a gap in case her son awoke from his sleep.

It amazed Christy that something as uncomplicated as water could wreak such havoc on heavy furniture like her refrigerator and bed. The latter had slid up against the doorway, pinning the door shut and was the cause of resistance when she had tried to open it. Rebecca's bassinet lay on its side, the white cloth soggy and dull from the nasty water. Her bureau was still where she had left it, more or less, but she knew that the clothes within were dripping wet.

"Miss, let's go," one of the men called out impatiently. She turned back towards the door and retrieved her handbag from a hook that was screwed into the wood. She had resorted to keeping it there to prevent Ryan from reaching it. He had discovered her secret stash of gum and breath-mints and raided her bag daily for his share. His little sweet-tooth had ensured that her wallet and other personal items were now safe and dry, high above the floods that had penetrated her home.

She quickly moved to her small closet to retrieve some dry clothes for herself and her son. He was still damp from his fall on the floor earlier when he had slipped in the puddle of water. It seemed like hours ago, but realistically no more than thirty minutes had passed.

Because of the cramped quarters that she shared with her children, her closet was divided into sections, which separated her things from her children's. She had the right half for her clothes and the left side was split between Ryan and Rebecca. Shelving on their side served better than hangers and she was quickly able to pull several pieces from the cubbyholes after she had set the flashlight down to free up her hand. She retrieved a bag from the overhead ledge and stuffed everything into it.

From her side she grabbed a clean pair of jeans, a T-shirt and a sweater. . On second thought, she quickly removed the wet clothes she wore and threw on the dry set, tossing another pair of pants and shirt into the bag. She pushed her arms into the sweater to complete her outfit. Her shoes were all waterlogged and served no better purpose than the ones she now wore. Ryan's shoes were lined on one of the lower shelves and had suffered the same fate as hers; however, as they would eventually dry, she threw a pair into the side compartment of the bag away from the dry clothes. A wall of vinyl separated them, ensuring that the wetness would not seep through.

Retrieving the flashlight, Christy glanced around

quickly, desperate to think of something that she should take before she left. She eyed Rebecca's baby bag hanging on the back of her bedroom door and breathed a sigh of relief. She absolutely could not leave the house without that. It was kept fully stocked at all times, as she had learned that failure to prepare in advance always resulted in her forgetting something terribly important. Without fail, in her usual rush to leave the apartment with her children, she always left something behind; from extra diapers to formula mix to bibs. She had learned to keep Rebecca's bag stocked with all of her needs so that it was just a matter of grabbing and going.

At no time had this practice served her better than this very moment. Christy looped her free hand through the strap and lifted it from the hook. A head peeked around the corner, startling her, just as Ryan's crying reached her ears.

"Ma'am, we cannot stay here any longer," the man urged, "The wind is picking up outside and I don't want to chance waiting any longer."

"I'm so sorry," Christy apologized, "Can you take these bags, please?" She handed everything to him, retaining the flashlight, which she fanned around the room one last time. Satisfied that she had everything that she needed, she stepped into the hallway and towards her sobbing son.

Ryan's face was contorted with fear from the belief that she had left him. Big tears ran down his cheeks as he chewed on his fingers. His lips curled, exposing his baby teeth as the pitiful wails poured out of his little chest. As soon as he saw his mother, he squirmed free of the arms that held him and ran to her, wrapping his tiny arms around her leg in a bear grip.

Christy tousled his hair gently, and then reached to cup his chin, turning his face upwards towards her.

"Hey now, little man!" she said softly. "I will never leave you, you hear me?"

Ryan sniffled and averted his eyes to glance at the man who stood beside her, as if trying to determine whether or not he was friend or foe. Hot tears blinked from his lids as he met his mother's eyes again.

"I will always take care of you, okay?" she reassured him. "Okay, puppy?" she insisted, pressing for confirmation from him.

He simply nodded his head and tightened his hold on her leg, determined not to let her go again. Christy reached for his hand and pried it loose, retaining the grip within her own hand to reaffirm her words of comfort.

He allowed her to do this and squeezed his little fingers around her bigger ones. She looked up at the two men who stood in her soggy living room and nodded.

The five of them moved close to the door, but were stopped short of exiting into the stormy madness. The window man put his hand on the knob but did not turn it; instead he turned around to face Christy.

"One of us will have to carry the boy," he said, looking down at Ryan. "The wind is still much too powerful and he will not be able to make it on his own feet. It will be difficult for us to do this, but we have to get out of here." He glanced up toward the kitchen ceiling, training his light on the spot the sheetrock had fallen from. Water poured down onto the countertop, unrestrained.

"I suggest you hold that baby tight and zip your sweater up over her," he advised. "Jim, you grab the boy, I'll hold onto the mother," he said, addressing his friend, the doorman. "We'll get out of the doorway quickly and hug the wall to the right. Then we'll huddle together tight-like and hurry across the road back to your apartment." He pointed in the direction of Jim's apartment.

Christy realized that they had come over from the new four-plex across the road from her. It was about six months old and featured two levels of apartments in a trendy design that caught her eye every time she drove into her little parking lot. She had often wondered what the inside looked like.

She quickly positioned Rebecca inside the sweater and gratefully accepted Jim's assistance with pulling the zipper together and zipping it up until it just covered the baby. She tucked the front part of the sweater into her jeans for added safety. Jim scooped Ryan up into his arms. The boy began to protest, but Christy soothed him with one hand on his arm.

"Remember what I said," she reminded him. Ryan settled down and everyone turned expectantly to the man at the door.

He slowly turned the knob and tested the pull on the door before opening it bit by bit. The cold air gushed into the room, blowing out the candle that sat on the shelf by the door. Hard rain slapped their faces, quickly saturating their hair. They followed the leader outside, careful to hug the wall as he instructed.

Christy turned her body into the wall, protecting Rebecca from the brunt of the assault. She zipped her sweater a little higher. Behind her, Jim sheltered Ryan in the same manner. The young child pressed his head deep into the man's chest, one hand tightly wrapped around Jim's neck and the other curled over his own head, shielding himself from the rain.

The flood had receded back into the sea, which was about half a mile away, but there was still significant water coverage everywhere. This area was not so low; however, higher ground would have settled vast amounts of water as well due to the continual downpour. Lightning flashed in the sky overhead like fireworks, brightening the day but not enough to dispel the gloominess. She saw the boat that her son had spotted earlier. It was now on its side, resting across the hood of her car.

The nameless window man carried his flashlight in his right hand and reached for her free hand with his left. He looped his arm behind her elbow, interlocking their limbs. She used her left arm to cradle the baby in her sweater, so was unable to do the same with Jim.

A tug on her arm indicated that they were stepping away from the building into the road. She looked back to ensure that her son was right behind her before taking the step into the unprotected expanse.

The gusts of wind beat Christy's body from every direction like a tag team of boxers pumping fists into her face, back and sides. She held her baby daughter tight against her chest as she took the pummeling.

Behind her, Jim and Ryan fought the attack as they all crossed the street, desperate to get back indoors where it was safe and calm. Her arm was linked firmly into the crook of the arm of the man who had peered into her window less than an hour ago. She still did not know his name, but was so grateful to him for saving her small family.

They snaked through the puddles and calf-high waters, bent almost at the waist at times to buttress themselves against the gales that catapulted raindrops at them like missiles. Christy was so afraid that she would stumble and drop her infant. She fought to make each step sure and steady, while simultaneously maintaining her balance against the blasts of wet air.

The group finally reached the building and huddled against the wall to catch their collective breaths. The building now stood between them and the force of the hurricane, shielding them from the apocalypse that was threatening to crush them. The howling continued but did not penetrate every range of their hearing. A slight stillness settled under the shelter that they had found.

The men moved off, signaling for Christy to follow. They walked past the first door and stopped in front of the next one, which bore a shiny brass '2' centered at eye level. Jim's knock was answered by a woman who quickly ushered them inside and shut the door behind them with a forceful push, returning several sandbags against it. She had big, plush beach towels on hand and proffered them to each of the soaking adults. For Ryan, she procured a

HURRICANE IVAN: THE EXPERIENCE

smaller towel with dolphins happily jumping across its cresting waves. Jim set his down and began to dry him first.

"Oh my Lord!" she exclaimed, clutching her chest. "My heart has been in my throat ever since you both left! Hi, I'm Janet and you've met Jim and Keith." She extended her hand towards Christy.

Christy had unzipped her sweater and released a crying Rebecca from her warm cocoon, setting her down on the nearby sofa. She quieted her with the remainder of her bottle before turning to Janet, eyes full of tears and relief that they had arrived safely. Words escaped her as she reached forward and hugged the woman, and then Jim and lastly the window man, who was now named Keith.

"How did you know to come and get me?" Christy asked.

Janet smiled and looked at Keith. "He's been looking out of the window all night," she said, winking at him.

"Well, we saw you earlier in the day yesterday and figured that you were staying home through the hurricane, so I just wanted to make sure everything was okay," Keith answered, shooting Janet a look intended to silence her. "When the lights went out last night, we saw the candles in your window and then when the water started to rise this morning we decided that it was probably not going well in there for you. There was nothing we could do until it started to recede, and then we just bolted for your place."

Christy turned to glance at the window, wondering how they had watched her place through the hurricane shutters, but saw none. She was able to see right through them like a giant television screen that channeled some freakish weather program. The grayness coupled with the wind-driven rain was like a surreal movie that she had seen somewhere.

"Aren't you worried that those windows will get blown out or hit by something and shatter?" she asked.

"Naw, those are hurricane resistant," Jim announced proudly. "We also have flood-proof doors, but Janet insisted on getting sandbags too. They really built this place to stand up to ol' Ivan!"

"C'mon, let's get you guys dried off and into some warm clothes," Janet said. "My son is about the same size as your little man, so I can get some of his clothes for you. He has an extra bed in his room so feel free to take that, 'though I don't think the three of you can quite fit on it. Maybe we can set something up on the floor for him, and you and the baby take the bed."

Christy nodded, just happy to be in a place that was hurricane proof and dry. Janet led the way with a lantern illuminating the hallway in front of her, although it was not entirely necessary. Christy picked her baby up, careful not to hold her too close to her wet clothes. She took Ryan by the hand as he let out a long, wide yawn. She could not help herself from doing the same. Yesterday had been a long day and an even longer night as she had sat alone with her children, fretting and worrying about the storm. The little sleep she did get in the early morning hours was not enough to grant her complete rest. She could not wait to curl up and fall back to sleep, knowing that she stood a better chance for survival here. She was relieved to be in the company of people who were so kind and gracious. Night would come so quickly, and after the recent mind-numbing events, who knew what it would bring?

SEVENTEEN

Buckets now lined the hallway and towels were laid to catch the splashes as the rains continued to infiltrate the apartment from the damaged roof. Des had appropriated a 55-gallon drum from the backyard and positioned it strategically beneath the largest leak. Patty worked tirelessly with the mop, not allowing water to stay on the tiles long enough to take up residence or travel in directions that led out of the hallway and into other rooms. Jackie and Michelle occupied the children with snacks and games to keep them out of the way and entertained.

Predictions had been made that the hurricane would pass over by two o'clock that afternoon; however, at two-thirty it seemed that there had been little or no change to its intensity. Des sat at the window looking out through the opening that was not completely covered by plywood. With no radio, they were unaware of what was happening to the rest of the island. The weight of his mother, grandmother and aunt's fate rested heavily on him. Since the phone call was cut off earlier that morning, he had not been able to reach them. His phone was permanently fixed to the palm of his hand, and he was unconsciously hitting redial every few minutes.

Whenever his sons entered the room, he tried to appear unconcerned and cheerful. They were his life and he would avoid causing them stress at all costs.

"I think that it must be about time for a nap, what do you think?" he announced, as Demitri crossed to retrieve a ball that had rolled into the room.

"No! I don't want to take a nap," he protested, "I want my Gela!" He grabbed the ball and retreated to Naldo's bedroom. Michelle wanted Gela too. She had been their nanny since the boys were 12 days old and was ingrained in the family unit. It was so convenient having someone that she could depend on to help with the enormous job of raising the boys. Michelle wondered how Gela was holding up in her little house a few miles away.

"A nap is not a bad idea," Michelle agreed with a weary smile. "I have run out of things for them to do and they are eating everything in the house. A nap could last up to three hours and would give us all a well needed break." She headed for the room to make the decree amidst loud objections.

Des turned back to the window and waited for the wind to die. "What is going on out there?" he pondered under his breath. His question went unanswered, as the storm did not yield to him but rather continued on in its destructive path.

EIGHTEEN

September 12th, 2004
1500 hours
Ship call sign: P3JA8
Hurricane Report: Latitude 17.7 Longitude 81.8
Approximately 190 miles SSE of Grand Cayman
Category: 5

The children had been asleep for over an hour when the adults noticed the relenting of the wind outside. Des looked at Michelle expectantly, for affirmation that they would chance a hike to their home to assess the damage.

Jackie agreed to keep an eye on the children while they were gone. She knew that they would be no trouble and understood that both Des and Michelle needed to see the condition of their home or risk grooving her tiles from pacing back and forth with worry.

"You two go on! They'll be fine – you both just be careful!" she warned. "Bring all the towels and buckets you can grab, though!"

"We won't be gone long," Des assured her. "Just long enough to assess any damage and try to contain it." He looked at Michelle, his eyes betraying his panic. She winked at him and turned to go out through the garage. To the left of the electronic door was a small aluminum door that led outside. There were no shutters or plywood boards on this one. It rattled against its casing, echoing throughout the hollow room. Its narrow window measured two feet wide by three feet tall, and the movable louver was slightly opened to alleviate the pressure that had built up from the hurricane. Rain splashed in through the gap and formed puddles on the concrete floor. The dogs were busily lapping it up as Michelle slowly put her hand on the doorknob.

"Wait!" Des blurted, before she could turn the handle. "I don't think it's a good time to go yet."

"Yes it is," she answered him, irritated that he was stalling. "The wind has died down enough that we can run home. It's not blowing hurricane force anymore and it's late. If we wait too long, it'll be dark and with no electricity, we won't see the house until tomorrow. I can't wait that long!" She proceeded to grip the handle and turn counterclockwise to release the latch from its alcove.

"But if we wait a few more minutes, it might calm down even more and we won't risk the danger of getting hit in the head by a flying shingle."

"Well, if you want to wait around until then, that's fine by me. I am going home right now, so if you're coming, let's go; if not, I'll see you when you get there!"

With that Michelle gave a final twist on the handle, simultaneously pushing against the door with her hip. The door was pinned by the wind for a moment, and then suddenly ripped out of her hand and flung itself outward, crashing against the side of the house. Michelle was sucked out of the garage by the abrupt impetus, but remained on her feet as she struggled to shield herself from the wind and rain. Her feet bore no protection other than her rubber flip-flops, which quickly bogged down into the saturated earth, creating deep suction noises with every footstep.

Des was at her side in an instant, drawing her closer to the house. The roar of the wind was deafening and they had to shout to be heard above it. Des looked down at her slippers disapprovingly.

"Why did you wear those?" he yelled. "They are only going to slow you down and cause you to slip and fall." The water poured down his nose and cheeks as the rain swarmed him like mad bees. The taste was both sweet and salty, evidence that the winds were driving the sea upwards and spraying the land with a mixture of cloudburst and ocean liquids.

"It's okay, I'll keep up," Michelle replied. She could have easily worn her tennis shoes, but thought these shoes would do, as their home was only a short way up the road. Michelle peeked around the side of the wall that shielded them from the full brunt of the currents of air. She wore her hair in a ponytail to avoid being blinded by the lashing of her long mane, but even now the loose wisps were assaulting her cheeks and neck as the gusts teased them into a frenzy. She pulled away from the corner and looked at Des.

"On the count of three," she shouted, "we run out of the gate and through the neighbor's yard into ours. That's the quickest route."

"Fine." Des agreed. "One…two…" Michelle bolted from their nook and was immediately swept along, the explosive gusts mixed with rain, pasting her shirt and shorts to her back and rear. Des jumped from his safe area and too was caught in the draft. They ran to the gate and fiddled with the latch until it gave way, releasing them into the lane beyond. Michelle suddenly thought of Dorothy in *The Wizard of Oz* as she tried to get home before the tornado, shouting "Auntie Em! Auntie Em!" She was tempted to look behind her to make sure that the Wicked Witch of the East was not cycling insanely behind her, cackling and peddling midair in the storm.

Suddenly, her foot disappeared into a hole filled with water. She halted so abruptly that Des did not see her stop. She struggled to pull her foot out as the wet ground beneath enveloped her slipper in its murky grasp. Finally her foot tugged free, leaving the shoe behind, but she did not hesitate to retrieve it. Instead, she ducked her head down against the wind and looked for her husband's figure. Des was nowhere to be seen, so she assumed that he had already made it the house and was safely inside.

She continued on the route that they had agreed upon, crossing through the front portion of land that housed apartments adjoined to their property. At the edge, she noticed the first sight that ripped her heart. The 100-year-old Tamarind tree that stood at the corner of their land was toppled over, its roots stretching about 30 feet in the air. She paused in front of the apartments to catch her breath and behold the awesome spectacle before her. That tree had provided sweet fruit when in season, as well as immeasurable shade in the heat of the day. Its branches had spanned across the area where they parked their vehicles and were intertwined with the branches of another tree that grew at the other corner. The Fichus tree was also felled, but that one lay in such a way that it blocked the path that led to the main road from Jackie's home.

"Come on, hurry!" Des shouted from the middle of their back yard. He stood hunched over, beckoning to her with urgency. Michelle picked out a path quickly and negotiated against the wind and rain to join him. Stray roots and rocks tortured her bare foot as she ran full tilt toward her husband. His outstretched arms pulled her into him as they navigated the final steps to the back door through the fallen branches of the Tamarind tree, which extended in a death stretch to the back of the house, only inches from the wall.

As he fumbled with the keys, Michelle turned to survey the yard. Trees lay everywhere, nothing remained standing. The sea of elevated greens and dark browns fluttered and swayed as the wind tried to revive them in an artificial respiratory dance.

Des finally found the right key and inserted it into the lock. He tugged on the door, which seemed swollen and stuck in the frame. Michelle put her hands around his and used her shod foot to brace against the wall. They both heaved and the door gave a bit but stubbornly refused to open.

"Pass me that bridle over there," Des shouted, as he pointed to the burglar bars that sealed the kitchen window from the prying hands of would-be thieves. Wrapped tightly around one of the bars was a horse bridle that he had obviously chosen not to move. His decision to leave it right there was disclosed by the neat and deliberate manner in which the equipment was secured to the steel bar. Michelle removed it and passed it to him with a frown visible on her face.

"What?" he questioned innocently, not looking her in the eye. He took it from her and wrapped it around the doorknob, bracing his leg against the wall for more leverage. One solid heave freed the door, which he had to shoulder open against the force of the winds. He held it for Michelle to enter and then quickly followed as the door crashed shut behind him.

The air inside was musty, the dankness infiltrating their nostrils. It was dark inside, as all shutters had been lowered and all windows without them had been covered with plywood. Des quickly retrieved the flashlight from the laundry room and turned it on. They expected water on the floor, yet their shoes trod on dry tiles. They had anticipated the sheetrock ceiling would be soaked, dripping water everywhere, perhaps even fallen in some areas. That was not the case however, as Des shone the beam of light across every inch of it.

Both Des and Michelle held their breath as they examined room by room, preparing for the worst as they turned the doorknob to each section of the house. The final door led to their bedroom and they paused, afraid that the horrific sight that might lay before them would dispel their mounting excitement. Slowly Des turned the handle. Michelle stood behind him, gripping the hem of his shirt, her fists clenched and pushing into the small of his back. He released the knob when it disengaged, and pushed the door open. They glanced around from wall to wall, surveying every square inch from floor to ceiling. A small puddle of water pooled near a window that allowed daylight to pour in. It had been covered when they left, but somehow the wind had ripped the shutter in two and was slamming it mercilessly against the wall outside. The window had been cracked and rain was seeping into the room and slowing running down the wall.

Michelle grabbed a towel from the adjoining bathroom and tossed it on the floor to soak up the water. Des completed his assessment of the room and suddenly threw his hands in the air.

"Praise God! Thank you God!" he shouted, dancing around the room. Both he and Michelle went room to room once more, yelling praises and thanksgiving as they realized that their home was truly spared and there was no more damage than a loose shutter and a cracked window.

Des grabbed Michelle's hand and burst into spontaneous prayer. He gave thanks for his family's well being and for the protection of his home from the storm. He prayed with fervor, lifting up exaltation and worship. Time was of no consequence as they sought to honor their God with devotion and blessings.

Their hearts were full and the tears that flowed

freely were of sheer joy and relief. Des and Michelle hugged, their beaming smiles exhibiting the true elation that could only be voiced through their prayers of adulation. Their spirits lifted high on the wings of gratitude where words became useless and attempts to express relief were met with more tears and wide smiles.

Outside, the winds continued to howl, but a new sound joined in that bore a likeness of the airy baying of the storm, but had a distinctive animalistic quality to it. Suddenly Des realized that the sounds were coming from the front porch where his dogs were being sheltered. He quickly ran to the front door and unbolted the lock. The sandbags his mother made for him were stacked against the base of the door. They were soaked with an acrid mixture of rain and canine fluids. His dogs clawed at the door outside, desperate to get inside with him. As with Jackie's dogs, he was careful to cover the doorway with his body in order to prevent their escape from their makeshift kennel.

While Des attended to his dogs, Michelle hurriedly gathered towels and canned food as well as several changes of clothes for all four of them. She pulled out two large duffel bags and began to stuff them. Her mind mulled over the mental list of items that they would need over the next day or so if they continued to stay with Jackie and her family. She wanted to ensure that they had enough food and adequate entertainment for the kids. Naldo was well stocked with games and toys, so that would not be her main focus. Living on an island meant that one always owned at least two beach towels per person in the household; therefore, Michelle had no problem locating several for her stash. She hurriedly moved around the house, gathering her bounty and tucking everything into her bags.

Des was met by his two extremely excited pit bulls on the front porch. The mother and son canines tag-teamed him with nudges and nuzzles that were intended to express their urgent desire to get into the house and away from the deafening roar of the monster that lay beyond the threshold of the small enclosure. They paced back and forth, jumping up on his legs and pushing off, marking his wet legs with an unsavory blend of mud, urine and other toxic waste.

"Easy now," Des cooed as he pushed the mother away and shielded himself from an attempted jump from her son. "It's all right now, the worst is over."

He surveyed the porch and each animal for signs of damage or injury, but found none. Their temporary shelter had proved safe and secure; his beloved pets were well.

NINETEEN

Dan finished cleaning up the broken glass that had littered the kitchen floor. He was not opposed to doing chores around the house, but Fiona was so much better at it than he. Right now, he just needed something to occupy his hands so that his mind could focus on a task rather than on the imminent destruction of his home and country. He was not usually such a negative person. This hurricane experience was really taking its toll on him. He had been through much worse in his life and thought he'd done just fine.

So why now was he such a basket case? How embarrassing that he would have such a sissy meltdown in front of Fiona's whole family! They were already skeptical of his marriage to her, given the age difference, and now they would just think he was some crazy old man who didn't have what was needed to properly care for her and protect her.

Dan watched her as she walked towards the bedrooms with a train of children behind her. She was such a mother already. She had ensured that they were all fed and was now leading them into the bedroom to prepare for an afternoon nap. He wondered how much sleep he would be permitted to enjoy tonight. The day had been one constant cleanup as water enveloped the house on the outside and found entry at every possible point on the inside. The storm had been raging for such a long time at high velocity for more hours than he could remember.

The other adults sat in the living room, discussing the events. Their hushed tones suggested that the conversation was not fit for young ears. Their worried faces shot frequent looks upwards as they surveyed the damage to the ceiling. He washed and dried his hands in the kitchen sink with a bottle of water that Fiona had put aside for that purpose. He momentarily lost himself in his thoughts as he watched the tiny bubbles multiply on his fingers from the friction he created as he lathered. His mind raced through the scenarios that might play out over the next few hours of light that they must endure before the sun once again set. It did not matter if there was light or dark outside, as the shutters over the windows ensured that they would always been plunged into darkness when the lanterns were turned off. His concern with nightfall was more to do with the potential need to evacuate should conditions inside his house become unbearable or dangerous.

Dan had lived through many hurricanes, but the majority had only affected these islands during daylight hours. They were never safe during such a ferocious storm, but at least when there was sun somewhere above the dark clouds, you stood a good chance of surviving, as long as you could see what was going on. Ivan had started to pummel them so early Saturday evening that he was sure the worst would arrive before this side of the world completed its 360 degree rotation and once again received the light of the sun. Now they were well into the middle of Sunday, and the hurricane was still blowing with enough force to roll cars down the street.

Without daylight, one's chances of successfully escaping a life-threatening situation were greatly reduced. The dangers unseen were far more sinister and harmful. As Dan pondered each one, Fiona slid up from behind and hugged him. She reached for the bottle of water and poured it over his soapy hands, rinsing them. He shook them in the sink to remove the excess water before he swiveled to accept her embrace front on, the water from his clean hands dripping onto her back and to the floor that he had just mopped.

"Well, the kids are all tucked into bed," she said, as she breathed a long sigh of exhaustion. "Tonight you can take the bottom bunk and I'll sleep on the couch."

"I'll be fine," Dan responded. "You have so many comforters in that hall closet that we can just make one big bed in the middle of the living room. So far, the roof in there is holding up pretty good. So as long as we keep the water on the floor under control, we should stay dry." Fiona smiled up at him, then released her hold and walked away in search of the bedding.

Dan joined the others in front of the blank screen of the television set. They were discussing how long the wind had been blowing outside and wondering how much of the hurricane already passed over and how much was left to come. Chances were good that it would continue into the night before they found relief, but without any connection to the outside world, there was just no telling. Dan wondered what tragedy awaited them on the other side of that hurricane. Grand Cayman was so small, though it was the largest of the three, that word travelled rather quickly. Bad news usually travelled even faster, so he was certain that once the 'all clear' was given and they were able to emerge from their shelter, reports would begin to pour in.

"Lord, help us," his mother-in-law whispered. She sat on the loveseat beside her mother, hands resting on her lap, wringing a handkerchief with much anxiety and consternation. The worry that etched her face concerned him all the more. These were women who had lived through decades of storms and were certainly well seasoned on what to expect. Their concern only served to heighten his own fears.

He shifted his eyes to observe Fiona's brothers. They were both tall, solidly built men. Dan got along with both of them well, despite their initial disapproval of his marriage. Each man sat hunched over on the edge of the sofa. Their forearms rested on their knees with hands tightly clenched, fingers interlocked as if holding something securely between their palms. Their faces were creased with fret as they listened intently to one another. . Ernest's wife, Julia, sat with her legs curled beneath her, one hand extended and resting on the back of her husband, while her other hand pulled and picked at a piece of string that hung from a button on her blouse. Her eyes were fixed on the door, unblinking and glassy.

The voices stopped suddenly as something hit the roof and was dragged a few feet before dislodging and resuming its flight. No one spoke, each waiting silently for something to happen above their heads, but only the blowing outside could be heard.

"We'd better empty the buckets and pots before we have several inches of water in this place," Dan announced as he rose to his feet. His voice broke through the meditative silence, snapping everyone back to reality.

Ernest stretched and yawned as he stood, leaving Julia's hand to fall to the sofa as it slid from his back. "Yep, yep, we need to keep on top of this t'ing," he said as he stepped past Dan.

Ray jumped up and headed toward the kitchen, clapping Dan on his back as he passed. He grabbed up the container that sat in the middle of the floor, collecting the drainage that had overtaken the light fixture, and dumped its contents into the sink. He quickly returned it to its designated spot, but not before a puddle had formed. The mop still leaned up against the cabinets, so he retrieved it and wiped it across the floor.

In the bedroom, Michelle pulled out her video camera and switched it on. The battery gauge showed that it was very low. She chided herself for not thinking to charge it up before the storm, when they had electricity.

With everything packed, they were ready to make the short jaunt back to their friend's house, but curiosity was pulling her toward the front of the house. All windows were boarded, and with the awnings drawn down and tied on the porch, she would have to exit through the back door and make her way around the side of the house to the front. She desperately wanted to see what lay out there.

"Des, I am going to film the backyard and see if I can make it to the front," she called through the front window.

"Give me a minute to finish up here with the dogs and I will come with you," he called back, but Michelle was already heading toward the back of the house. He heard her throwing her body against the back door, trying to get it to open, so he quickly petted the dogs and went inside.

Just as he reached her side, the door gave way, but barely opened an inch before the wind fiercely slammed it back on them. Exchanging looks and places, Des tried his luck and was able not only to push the door open, but also to hold it in place so that they could peer into the backyard without being assaulted by the whips of the wet wind.

"Let's go to the left and come around on that side of the house," Michelle suggested as she removed the lens cap from the video camera. She pressed the button to record and began sweeping from left to right in a slow movement, ensuring that she captured every fallen tree. The camera had a small viewfinder that opened and extended from the side, that allowed her to film without holding the equipment up to her face. She was able to keep it at a distance and look around to see where she wanted to point it. As she turned, she noticed a wooden structure that looked familiar, but oddly out of place near to the back door. She panned the camera in that direction and zoomed for a better look. Suddenly, movement enlivened the screen of the viewfinder, and she realized that she was looking at the overturned rabbit cage that housed the family pets. She nudged Des and pointed to the cage, making sure that she did not lose her visual with the camera.

"I can't be sure, but it looks like there are a few that aren't moving," she said as she played with the zoom, trying to focus in on the contents of the overturned bunny housing. "I told you that you should have brought that cage closer to the house and maybe tie it to something bigger."

"It was just too heavy to move by myself," Des explained, sorry that he had not called on a friend to assist with the move. Once again, Michelle had been right, and this delay had resulted in death. He felt badly.

"I am ready to move, are you coming?" Michelle asked as she stepped out of the doorway, but remained in the shelter of the door that Des still held against the constant push of the wind.

"Yep, let's do it," he said, placing his foot on the ground beside hers.

He turned to hold the door with two hands, and had to use the weight of his body to prevent it from slamming back into its frame. Securing it just enough to keep it in place without letting it wedge, he released the handle and grabbed Michelle's hand as the wind shuttled them to the corner of the house. The view up the side of the house was both interesting and more of the same. Renegade branches lay strewn along the pathway, having abandoned their trees for more secure housing on the ground. A dead chicken lay twisted, its neck obviously broken, under a branch from the almond tree that once shaded the western quadrant of the front yard. The wind gusted through this little alleyway with less force, yet packed enough power to drive them to the walls, which they hugged as they made their way forward.

Over the fence that ran parallel to their home stood another house, occupied by a neighboring family. Michelle could see movement in the only room that was not hidden behind a piece of plywood. She called to the occupants and was relieved when a head pushed up to the screen and waved back. Words were lost in the cacophony of sounds. They moved onwards, feeling a sudden change in the pressure as they neared the front. It was almost like a vortex, pushing and pulling the closer they got. The wall gave way to an awning that was drawn over the front porch. Des knew that his dogs were on the other side, listening to everything and no doubt could detect with amazing precision the sound of his voice above the wailing and moaning of this wind.

Michelle continued to film and attempted to narrate, not sure if the playback would produce discernible words with the video. As she neared the end of the porch, she noticed that there were more branches lying horizontally on the ground, evidence of another fallen tree. She tried to move forward to get footage, but was pounded backward by a sudden gust of wind. Des stood behind her and was able to fortify her position by bracing her back with his palms.

Panning the camera, she noted that the tall Poinciana tree that had provided enormous coverage to the front yard was lying completely on its side, parallel to the house. Not one branch came within touching distance of the house. The yard itself was strewn with every material imaginable. The beach sand that was the basis of the old fashioned yard was barely visible under shingles and leaves. The wind and rain thrashed everything about as if the rinse cycle was in permanent motion.

"It's such a mess," Michelle mumbled half to herself, knowing that being this deep into the storm would make it hard for the camera to pick up her voice, so shouting would just add to the accumulating loudness.

Des remained behind her, holding her steady as the wind persisted in its efforts to drive her backward and into the house. Its molesting paws writhed viciously over her clothes and hair, assaulting her with grains of sand and pellets of rainwater.

Standing beside the porch, she could smell the dogs and their waste. She heard the whimpers and yelps as they pleaded with her and Des to come back to safety and release them into the wind. They longed for the freedom to run and frolic as before. This prison of filth and darkness was not for them.

Hopefully they would get their wish sooner than later. God only knew how long this would take to blow over.

TWENTY

Des felt the wind's power buck against Michelle with every burst. He held his right leg behind him and locked the knee to fortify his stance as he pushed against her, his left leg bent in front. With each blast, his elbows threatened to buckle.

He had never experienced anything like this in his life. He tried to get a good look at the front of the house, but the rain was so hard and piercing. He closed his eyes to protect them from the sting. He would just have to watch it on the video that she was making. His clothes clung to his body as the heavens wrung out each cloud overhead. The salty taste seeped between his closed lips without permission. Was she almost done?

He peeped through a lowered eyelid to glance ahead briefly and caught sight of the felled tree by the road. Now that it was gone, he felt a twinge of regret. In all the years that he had lived in this house, practically all of his life, he could not recall much time under that tree. Of course, he must have enjoyed its shade at one time or another, but he felt that he had taken its presence for granted all this time. Its large trunk curled up from the ground where its roots still clung to the depths of the earth, refusing to let go of the nutrient-rich soil that it needed to survive. The tree was partially covering the main street that ran through this part of the district. Traffic would be slowed considerably at this point. How many times had he even climbed it as a boy? Now his own children would never have the experience.

A sudden gust threw more rain and sand into Des's face, forcing him to tuck his head deep into his underarm, releasing one hand from Michelle's back to instinctively protect his face. Michelle's sudden loss of footing caused them both to stumble backwards. He grabbed her hand and began pulling her towards the back of the house as she tried to resist. He persisted with a tug and she acquiesced, letting him lead her to shelter.

Back inside, she dried off the camera and placed it carefully inside its casing, eager to show the kids and their friends a different view of the storm. They quickly changed into dry clothes and made the short trek back to Jackie's house with their loot.

Relieved that their house and dogs were okay, Des settled more comfortably with his family. He sadly relayed the news of the demise of some of the pet rabbits to his sons. They took it well, he thought.

He had still not heard from his mother or aunt, and that worry ate at his nerve endings like slowly eroding acid. He replayed the last conversation over and over, trying to pick out anything that would calm him. Nothing did. He needed to get there – it was that simple. As soon as the winds died down some more, he would be on the road.

Christy awoke in a strange room with small airplanes dangling from the ceiling. Her mind was fuzzy and she struggled to remember where she was. Gray light poured into the room from a single window that was thinly covered by a curtain, bearing more airplanes captured midflight in its folds of fabric.

She glanced at her wristwatch, which told her it was just past three o'clock, and rubbed her eyes to wipe away the blurriness. Her elbow hit something hard beneath the covers. She peeled them back to reveal Ryan, who was snuggled up against her like a kitten, curled into ball and pressed against her body. Christy looked beyond Ryan for her daughter, but saw only carpet. She sat up with a jolt and flipped around to see if Rebecca had been behind her, but was met with the edge of a bed. Panic gripped her tightly like a vise squeezing the air from her lungs. She jumped to her feet, all sleep purged from her in one instant flush.

She quickly threw back the covers of the makeshift bed that had been made up for her on the floor between two twin beds. Ryan's body was the only one there. He protested the removal of his covers with a groan, and rolled over before sinking back into slumber.

Christy laid the sheet back over him and turned to survey the beds. Both were empty; one was perfectly made and the other bore disheveled evidence that it had been occupied at some point, but not by her baby. She mentally returned to the events before falling asleep. Had she left Rebecca in her apartment? Did she drop the baby in the storm? Her heart raced; a hot flush enveloped her, coaxing an instant sweat to release from her pores.

Noises floated into the room through the closed door. She detected laughter, but was not sure. She hurried to the door and then towards the sounds down the hall.

"Good afternoon sleepy head," Janet chirped. "We were wondering when you'd wake up."

She held a blanketed bundle over her shoulder, gently patting and bouncing.

"I hope you don't mind me taking your little girl for you. She was crying a few hours ago, but you were so dead to the world I didn't have the heart to wake you. Jim showed me the baby bag and I found everything I needed in there to sort her out."

Christy let out an explosive sigh, not realizing that she had been holding her breath since she had left the bedroom. She smiled at Janet and said, "No, of course not. Thank you! I was just worried for a minute when I woke up and couldn't find her."

"You must be hungry," a man's voice said from somewhere out of her line of vision. Christy stepped out of the hallway and into the living room. She turned in the direction of the voice and saw Keith standing behind a counter in a large kitchen. Jim peeked over the top of the open refrigerator door and waved.

"You missed breakfast…which was actually brunch by the time we all got around to making anything …so you missed lunch too." Keith smiled. "I'm making some tuna sandwiches, do you want one?"

Christy nodded. Janet beckoned her over to where

she sat and patted the sofa. A little boy about the age of Ryan sat on the floor at her feet, playing with his toy cars. He looked up momentarily as Christy stepped past him, but quickly returned his attention to the imaginary car race that was obviously underway.

"Is it over?" Christy asked as she took the seat offered her. She reclined to look at Rebecca's face, which was soft and expressionless in her perfect sleep.

"No, he's still out there, blowing everything around," Janet answered. "I didn't know that they took this long, but then, this is our first real life hurricane, so I wouldn't really know."

"Not usually," Christy said, turning to look out of the window behind her. Daylight was more like twilight. The gray sky churned overhead like a motor. The clouds puffed out rain in heavy drops that gusted back and forth in a game of turbulent tennis. Christy was not sure she liked being able to see everything. At home, she had been behind the protective shield of the plywood, which, coupled with the darkness of night, had also reduced her ability to see. Now she had a complete view of everything that was going on outside, and it tied her stomach into a knot.

She focused on the small building across the street. Her apartment was taking a brutal beating. Most of the shingles had been stripped off of the roof by the angry hands of the wind. Tarpaper flapped like black sheets before ripping away and flying through the air. Some caught on poles, pasted against the wood, unable to move further. Other pieces were taken high above the ground, teased upwards by Ivan's fury.

Bald spots spread on the roof like brown cancerous patches. The wooden roofing sheets were being exposed as more and more of the protective covering was loosened and removed. Christy knew that nothing remained intact within her home. Wind and water would seek out every weakness of the building and enter to wreak destruction and obliteration of whatever it came into contact with.

She turned from the window, unable to watch any longer. Ryan stood in the hallway, rubbing his eyes. His hair was matted in every direction, and suddenly Christy was conscious of her own appearance. She jumped up and hurried to him.

"Bathroom's the second door on your left, extra toothbrushes in the cabinet under the sink and there's a gallon of fresh water on the floor that you can use to rinse," Janet called out. "If you need to flush, use the bottle of water on top of the tank."

Christy thanked her and led Ryan down the hall. At the mirror, she was appalled at the reflection that met her. Dark circles were crayoned under her eyes. Her hair stuck up a little in the back and creases on her left cheek suggested that she had not shifted her position the entire time she had slept.

She did her best to fix herself and then helped her son with his needs before they both returned to the others. Ryan quickly sat down with Janet's son, whom she had introduced as Matt, and was soon enveloped in the world of make believe.

Christy's tuna sandwich was waiting for her, along with an extra one for Ryan. She devoured it gratefully, feeling like someone who had not eaten in days. The sweet relish that had been mixed in was a welcomed ingredient that made it taste just right.

The group sat and talked well into the late afternoon. Christy shared her experiences from previous storms while everyone listened with rapt attention. The hours droned on as the weather outside persisted. The feeling of twilight persisted throughout the day, but with the large unobstructed window behind them, there was little need to add additional lighting until day's end. Lanterns were activated when the weak daylight could no longer provide ample brightness.

Conversations were easy and flowing. Christy felt safe and accepted with the group and especially with Keith, who stared at her often. He appeared to be in his mid-twenties, which was fine with her. She repressed a chuckle at the thought of finding love in the middle of a devastating hurricane.

At least she did not have to come up with a subtle way to tell him she had two kids. They had bypassed that obstacle at hello! For goodness' sake, the man had saved their lives. If that wasn't a good sign that he liked kids, then nothing was.

Janet played with Rebecca while her husband attended to dinner. Theirs appeared to be a loving and equal relationship. Christy admired them for their normalcy, and coveted it. This was a new experience for her to watch a couple this close and personal. All familial examples she had been exposed to were abusive and dysfunctional, which did not help mold her own. She knew that there was something better out there for her, but felt that her bad choices had marked her as damaged goods. She now had a lot of personal baggage that was surely a deterrent for Mr. Right.

"Can I help you with dinner?" she asked Jim. "I feel like such a burden on you, the least I can do is help in the kitchen." She rose from the sofa and walked towards him. Keith immediately followed her.

"Only if you're good at opening cans." Jim laughed. "No electricity, no stove. We're just having cold ravioli or soup tonight."

"I happen to be a very good can opener," Christy joked. "No batteries required!"

Jim handed her the cans and instructed Keith to grab the bowls from the cabinet. The kids were called to the table, which Jim had set with plastic forks and paper napkins. Janet insisted that everyone eat while she held the baby.

"You're spoiling me, Janet!" Christy exclaimed happily. "I don't think I have ever sat through a meal without a baby on my shoulder or a child on my knee. I could certainly get used to this!" She stole a glance at Keith, who was helping Ryan stab one of the meat-filled pastas. The sight of another man helping her child took her by surprise. She did not expect the sudden feeling of emotion that welled within her. It was scary and exciting.

"I don't know how you do it, Christy," said Janet, cooing to Rebecca between sentences. "I only have one, and some days I am so worn out. Jim can tell you! I am always asking him to take Matt so I can get things done."

"God never gives you more than you can handle

at a time, I guess," Christy replied soberly. "I know there have been many days that I didn't think I would make it, but by His grace, here I am. Safe, warm, and eating the most delicious cold ravioli I have ever tasted."

"Well, young lady, I am impressed. You have two beautiful children and your little boy is so well mannered and respectful," Jim chimed in. "Hey Matt, can you say 'Yes sir' like Ryan?"

Matt looked up, tomato sauce smeared over his mouth and chin. "Uh huh," he mumbled through a mouthful of food before looking back down for the next bite. The adults broke into laughter, which infected the two little boys, who also joined in.

Outside, Ivan was not laughing. He was angry, swollen, and unsatisfied. Nature was his tool and he used her to beat the island mercilessly. He commandeered everything in her arsenal and viciously pummeled houses and trees, boats and cars. Nothing escaped him that wasn't tightly secured or firmly tucked.

He raged as the hidden sun set on day two, and welcomed the night with fire in the sky and squalls on the horizon.

TWENTY-ONE

Sunday night was no better than the one before it. Sleep was a luxury they could not afford to enjoy at long intervals, for fear that when they awoke from their dreams, the house would be under water.

As had been the arrangement of the previous evening, the children were confined to one of the two bedrooms that had miraculously sustained little damage. Water damage could be seen in small patches across the textured ceiling. Julia monitored the beads of water in the glow of the lanterns until her eyes closed with sleep. Ernest checked on the leaks, often changing out the buckets that caught the dribbling flow before the water could crest the rim and overflow onto the tiled floor.

He considered turning the battery-powered lamp off before returning to the others, but decided to leave it burning and set it on the floor near the closet. It cast a band of illumination upwards without disturbing his slumbering family.

In the living room, his sister, Fiona, and her husband, Dan, sat curled up together on the couch, all transgressions forgiven. They cuddled their baby boy between them, making faces and talking to him with wide eyes and big smiles.

His mother and grandmother were on the other sofa, chatting quietly. Several hurricane lanterns lit up the room, exposing the gaping holes that dotted the ceiling where water had overpowered the flimsy covering and forced it from its perch. A large kitchen garbage bin sat in between the sofas, collecting the heavy stream as it passed from rooftop to rafter to earth.

"Where's Ray?" he asked, looking around for his brother.

"He went to rest a little on dey bunk bed," Fiona said without looking up from Evan. She bent forward to kiss his tiny toes.

"That's not a bad idea," Ernest said. "I'll just jump on the top one and hope for the best. Call me if you need me." He knew that he'd get very little sleep again tonight. Sheetrock continued to fall around them and new leaks sprang up every hour.

The initial shock had been replaced by passive acceptance of the situation. There was no getting out of it. They were boxed into this drama until the hurricane passed. The main goal was to make it through in one piece. The house could be fixed. He would help Fiona as best he could when this was over. He had not even allowed himself to wonder about his own house. There was more than enough stress with surviving this chaos in the here and now. He would deal with his own circumstances when he was able to walk outside and feel the sunshine on his face.

In the bedroom, Ray was snoring loudly, face down in the pillow. His legs were splayed out, one hanging off of the bed as if he had simply let himself fall carelessly to the bed no regard for how he landed. Sleep had obviously come to him before his head hit the pillow.

The king mattress still lay in the middle of the floor between the bunk and the crib. Dan's mother was curled up at the edge, curlers in her hair. Ernest stopped to do a double-take. Had she really taken the time to roll her hair? He laughed softly. Everyone had their own simple pleasures that helped them deal with tense circumstances. Humans needed familiar things to maintain their calm and sanity when trying times interrupted their routines and norms.

Without the chores or tasks that they were accustomed to, edginess would set in and threaten to steal what little balance they had left.

He carefully climbed to the top of the wooden bunk and took in the ceiling that his head was only inches from touching. Surprisingly, not a single drop of water presented itself for inspection. There was something about this quadrant of the house that had been built better than the rest. Not knowing what that was, and moreover not having the capacity to follow that line of thought until it led to an answer, he allowed himself to fall backwards onto the pillow and into a dreamless sleep.

TWENTY-TWO

Al had retrieved the battery-powered radio from the kitchen so that they could listen to the reports on the local radio station. The announcer was a local celebrity on the radio scene and had talked the islands through many storms. His personal sacrifice kept him away from his family and on the air almost around the clock, from the beginning to the end. His deep voice was soothing and caressing as he took calls and read press releases. He admonished the public repeatedly to stay indoors until the national weather service issued the 'all clear'. Venturing out into the storm would prove fatal.

For the lucky ones whose cell phones still transmitted a signal, their calls were gladly taken and aired live. Many callers were begging for help as the waters surged into their homes. Their cries filled the airwaves as the helpless listeners imagined what they were enduring. They could only sit and wait. The broadcaster attempted to calm each of them, but nothing seemed to help. The terror in their voices was palpable and transferred a feeling of anxiety and fear to everyone who heard them.

Though Mama DeDe and her household were

safe, they could not help but feel the panic rising in their chests as they listened to the terror-filled calls from those unfortunate souls who were experiencing the same trauma that they had just survived. Mama DeDe prayed that the others would also pull through unscathed.

The hours wore on and all that they could do was sit and wait. The sun rose again, though its rays never penetrated the thick layers of cloud that Ivan pulled around himself as he stomped back and forth across the sky above them. The garage filled with a natural light so dim that they were still forced to use the lanterns.

Chris and Al retrieved lawn chairs that Mama DeDe had secured inside, and everyone took turns napping throughout the day. With the passage of the floods of the day before and all the stress that it brought, severe exhaustion weighed them down and all but stopped them in their tracks.

"What time is it now?" Nay-Nay asked. "It feels like this hurricane has been blowing for a week."

"It's going on three o'clock in the afternoon," Massoud answered, after consulting his watch. Fortunately he had worn his waterproof one. "If we consider that this started late Saturday night or early Sunday morning," he paused, doing the math in his head, "it would seem that approximately thirty hours have passed, give or take. Is that right?" He stopped again, calculating the hours backwards to the point that the first real gust had blown.

"In all my years, I have never seen a hurricane last that long!" Mama DeDe said in disbelief, "That's gotta be a record or something."

The intensity of the winds outside had subsided some time ago. Nothing moved any longer and there had certainly not been any projectile explosions against the house in quite some time. A heavy mugginess had settled on them like a blanket, suffocating and stifling.

"All clear!" Chloe shouted, turning the volume on the radio all the way up so that the adults could hear it too. "He just said that the Cayman Islands are being issued the 'all clear'!" She jumped up and ran to the doors that led to the backyard.

"Chloe, don't open those doors!" her mother shouted, as she too rose to her feet.

"...I repeat, the hurricane warning has been lifted and the 'all clear' has been given," the announcer blared, his voice reverberating around the enclosed walls of the garage. No one made any effort to turn the radio down. "But people, please use caution and common sense if you must leave your homes. The government has issued a request that everyone remain inside and not wander about just yet. The authorities need to assess the situation outside and determine the hazards."

"We couldn't go anywhere even if we wanted to!" Nay-Nay laughed as she reached over and lowered the volume on the broadcaster. "The cars were under water, and even though I don't drive, I know that means they're useless."

Lacy stood beside Chloe, looking though the glass of the French doors. She called to the others who quickly crowded around to see. Beyond the doorway that penned them into their makeshift confinement, the yard was a total wreck. It was not a very big backyard, but it certainly served its purpose more than a few times. Mama DeDe had entertained many a friend and family back there. She was quite the hostess and invitations to her parties were coveted.

Large chunks of roof lay scattered about, while trees and shrubs were randomly strewn around, their uprooted corpses stripped bare of all leaves. It was hardly recognizable back there.

Mama DeDe reached forward and slowly turned the deadbolt on the door, her breath catching in her chest. As it slid across the inner cavity of the opposing door, it made a heavy popping sound, indicating that it has completely retracted into its casing. She used her other hand to turn the handle on the door. There was a slight suction, which quickly released as she pushed her weight against the glass. She expected to hear the sound made when you opened a vacuum-packed container or new jar, but it never came. The door relented and swung outward, coming to rest against the outside wall.

Mama DeDe was the first to take a step out of the house into this new world that Ivan had left behind. He had come to destroy them, to flood them out and devour them, but he had failed. Here they were, alive. Whatever he had done to the rest of the island would be dealt with. For now, they just relished their lives and thanked God for sparing them.

"This is what Noah and his family must have felt like when that Ark finally landed on dry land!" Nay-Nay said as she breathed deeply, thick, humid air filling her lungs. "And the air probably smelled just like this with all of those stinking animals."

Chloe giggled as she wrapped her arms around her mother's waist. Lacy returned the hug and kissed her forehead. "I wonder what happened to all of the people that were swept away by the flood. The Bible doesn't say anything about Noah seeing them later on," Lacy mused.

"Okay, that's enough, I don't want to talk about drowned people," Mama DeDe said, turning to walk back inside. "We made it through and I hope that everyone else did too, but until we know for sure, let's just not discuss that." She bent down to pick up a container that had been tossed around earlier and took it outside.

Al and Chris joined her, and together they began to empty the garage of its contents piece by piece, removing everything that could be lifted, and stacking it all outside to be sorted later. Chloe and Nay-Nay busied themselves with arranging Beatty on a beach chair, propping her up with a pillow and elevating her legs before joining the rest of the group. The only thing they could do was pick up the pieces and try to rebuild.

Michelle and her family emerged from the shelter that had held them prisoner for the past 36 hours. They blinked in the glare of the sunless sky. The clouds were dense and gray, inhaling and exhaling bursts of blustery breath that sent dead leaves swirling around angrily. The ground was littered with more than organic debris. Black roofing shingles, pieces of broken wood, and guttering lay randomly strewn about the yard and road.

The smell of salt filled the air as the nearby sea continued to churn and spew, sending up a saline mist into the gusting zephyr.

The look of the land had changed since the last time any of them had been outside. Trees that had once stretched their long limbs to the sky in verdant, graceful worship now lay in horizontal death posture. Branches splayed several feet outward, as if grasping for an invisible hand that would never help them to regain their natural stature. The leaves were wilted from an overdose of abusive salt and wind. They drooped earthward, curled in to reveal the lesser green side that had never been graced with the sheen and brightness of their flipside.

Michelle was the first to speak, breaking the human silence, though she had to raise her voice to be heard above the din of the wind and waves.

"Thank God we made it through!" she exclaimed. Des put his arms around her and squeezed tightly. The boys raced outside, glad to be free of their confines. They needed to stretch their legs and set their hearts racing. Their parents cautioned them to be aware of the loose ground.

Suddenly, Des released Michelle and turned to go back inside.

"I need to get to my mother's house," he called back to her. "I still cannot reach anyone by phone up there."

"Wait for me, I'm coming too!" Michelle said, as

she ushered the boys back into the house amidst their protests. "Jackie, would you mind holding the boys again? I just can't let him go alone!"

"Of course, child." Jackie smiled. "Just be careful!"

They raced to their house once again to check on everything before venturing out in the post-storm haze. Hesitant to use their newer truck, Des instead grabbed the keys to the old Explorer that his mother had given him when she had purchased her new Avalanche. The former was a reliable old vehicle, and he would not be so concerned about running it through the salty sea water that would inevitably be all over the roads between here and his mother's house.

As they slowly drove away from the security of the buildings that had protected them, their eyes widened in shock at the sights that awaited them. The road was littered with broken branches, fallen trees, and furniture. People were wandering around in much the same awestruck daze. Des called to every one of them by name - these were his people and he was just glad to see them alive.

"Boy, I hear that there are 700 dead, ya know!" exclaimed one man, as he walked up to the driver's door and rested his arm on the open window frame.

"No man! You're kidding!" Des said as he braked but continued to look through the windshield ahead of him. "Where is this? In one place or just in total?" he asked.

"I hear that a shelter collapsed and killed everyone inside up in North Side or East End," the man explained. Michelle's heart sank. Some of her family was from North Side, but she doubted that any of them were in the shelter. Still, while the thought of this thing causing deaths had always been a very real threat, it just seemed unlikely that it could ever happen. Her mind raced with names of people who were from that part of the island. The suspense of finding out which persons were no longer alive made her abdomen tighten.

"Where you trying to get to, Dessie?"

"Gotta try to get to Snug Harbour to find my mother. I haven't heard from her since last night and I'm worried." Des cleared his throat loudly, causing Michelle to jerk her head in his direction. He was still scanning the landscape ahead, but she was certain that she saw the glint of a tear in his eye.

"Cha! You better save your gas, my friend," the man remarked. "No one can get past Governor's Harbour. The sea washed the beach clear 'cross the road over there. Keep trying to call her, nah. Anyways, I'd better keep going, I gotta get to Mount Pleasant to check on my people. Later." With that he shook Des's hand and pushed off the vehicle. Des released the brake and continued driving.

"Where are you going?" Michelle asked. "He said that the roads were blocked."

"I know. I just want to check on the horses while we are out here."

They made the turn onto Watercourse Road. Des continued to drive slowly, dodging debris in the road every few feet. The sounds of cracking limbs could be heard beneath the tires as he was forced to drive over the smaller pieces that posed less threat of damage to the Explorer.

Windows down, the air that rushed in was still forceful enough to blow Michelle's hair roughly about, as if Des were actually driving 40 miles an hour instead of five. At times, she had to blink rapidly to provoke her eyes to water and rinse the saltiness away that the wind inflicted on them.

TWENTY-THREE

Pulling up to the property that they owned and where he kept the horses, Des was a bundle of nerves. Hearing that so many people were dead, not knowing how many more would be found and not being able to reach his mother was tearing him up. He had so little control over what had happened and would happen.

He prayed that his horses were fine. That by some miracle they had survived this awful hurricane as he and his family had done.

At first, he was unable to see anything for the devastation that greeted him as he slowed to a stop at the gate. Massive trees were down, large sheets of corrugated zinc roofing were wrapped around light poles at various heights, and everything looked unfamiliar.

For a moment, there was nothing to be seen but destruction and devastation, but then, from the farthest reaches of the land flashed the welcome sight of a dark figure racing towards the fence line. Behind that one was another and then another. He counted them as they appeared. One, two, four...six, seven. All there, all alive!! He looked at Michelle, a wide smile stretched across his face. She was smiling back, and squeezed his hand in an affirmative gesture.

He jumped out of the Explorer and passed under two rows of the wire fencing that surrounded the property. The animals were alive, but spooked. He knew that it had been a rough 36 hours for them too. Out here in the open, their senses were super alert. These were flight animals, which meant that they slept lightly, on their feet, ready to bolt at the first sign or sound of danger. With the world seeming to crash down around them for that long, he doubted that any of them had rested at all. In fact, they would have been running and dodging all of the falling debris, trees and anything else that they feared during the entire ordeal. The poor beasts were as stressed out as they would ever be.

He touched each of them, careful not to add to their anxiety. Even though they all knew him, he was certain that they would not hesitate to kick him or rear up if they perceived him as a threat, and in their exhausted state of mind, misconceptions were easy to have - even for a horse.

"I am going to ride there!"

"What!" Michelle asked, not quite sure she had heard him correctly.

"I'll drop you off back at Jackie's and grab a saddle, then ride to Snug Harbour. With a horse, I can cross whatever terrain is out there. It'll take a little longer, but at least I'll get there." He crossed under the fence again and returned to the driver's seat, starting the engine.

Michelle looked at him for a moment and then

said, "I'm coming, too."

"Michelle, stay with the kids. You haven't ridden in years and these horses are not in the best shape to ride. I am way more experienced and would be too nervous making sure you are doing okay and trying to get to Mummy."

"Des, I want to come! I am just as worried about her as you are. I will be fine." she persisted.

He let out a long sigh. It was obvious that he was mulling over an argument to make her stay, but he knew he'd never win it - she was determined to go.

"Okay, fine." He conceded defeat on this one, again. He was happy for the company, however. On the short drive back to the house, they were enveloped in a silence that filled with outer noises. The waves continued to crash on the rocks of the shoreline and the wind blew in lesser howls, but was nonetheless audible. More people were walking the streets now and many waved to the vehicle as it slowly passed.

Michelle touched Des and silently pointed at homes that stood in various degrees of damage, one hand over her open mouth barely concealing her shock and horror. One house in particular looked as though it had been run through with a wrecking ball. The roof was gone - completely. All that stood were the red walls, painted to look like bricks.

Des silently thanked God again for sparing his family, his home, and his animals. Now he just needed to know that He had also spared his mother, aunt and grandmother, too!

Back at Jackie's, Michelle quickly explained what they had found out and what they were going to do next. She then packed a bag with some water and a few snacks, as well as a change of clothes for each of them in case anything happened.

The boys begged to go with them, but Des explained how dangerous it was and promised to let them come the next time they went, when it was safer. One last kiss on their little foreheads and they were out the door again.

TWENTY-FOUR

Another night filled with mopping and damage control left the younger adults of Fiona's household suffering from unimaginable fatigue. The vigil that they kept in pairs was served in shorter periods, as each one fought sleep and collapse.

Monday morning brought with it a renewed hope that this would be the day that they could break out of the dilapidated house that once stood firm and solid.

The children traveled in a group now, tightly huddled, stepping past puddles that awaited cleanup, and occasionally over clumps of soggy sheetrock that fell from above. They were not noisy and hyper as they had been in the beginning. They were bored and irritable, but sufficiently frightened so as not to whine and plead to go outside.

Fiona and Dan struggled to care for their infant who, like his older cousins, was well rested and busily kicking and cooing.

The roar of the wind could no longer be heard through the shutters. They did not shake and stutter against the walls as they had been doing over the last two days. Even the water had been stymied, as if a great faucet had been turned off and only the weak drippings from the pipeline were finishing the trek to the end of their journey.

Their watches indicated that it was already afternoon.

Ernest and Ray conspired near the door. Fiona thought it was a strange place to have a conversation, and it worried her as she envisioned the door flying open and sucking them both out, as if it were attached to a flying aircraft a mile above the earth.

Both men turned to the household with serious looks.

"We think one of us should check out what's happening outside," Ray announced suddenly. No one spoke as they processed the ramifications of such a foolhardy act. Ernest quickly stepped away from the door and addressed the family.

"It's been almost t'ree days, man," he argued. "We can't hear nothin' going on, the water's almost stopped and I'm just goin' crazy in this house!"

No one challenged him, as they felt the same way and also wanted to escape to whatever lay beyond that door.

Ray moved to stand in front of the portal that separated them from the unknown. He rested his hand over the knob and took a deep breath.

Julia gathered the kids and escorted them into the bedroom, more out of the need to isolate herself from the potential threat that lay beyond than to actually sequester them. Fiona followed her with Evan. They shut the door behind them and sat on the bed in prayer.

The elderly ladies remained on the sofa, too

curious to move. They shifted their bodies to facilitate a better view of the doorway.

Ray initiated the sequence of unbolting the locks and turning the door handle. He slowing pulled the door inward, remaining tucked behind the wooden shield until he was certain that nothing was airborne and aimed at him.

The cool wind gusted into the boxed room, refreshing the stale air with salty breaths. A light drizzle patted the tile, but quickly moved on as a flurry of air swept it back into the yard.

Ray pulled the door wide, pinning it against the wall, and stepped outside. The squall was subdued but still seesawed around the house.

Trees were strewn about, tangled with power lines that had fallen with the poles they were attached to. Leaves of every size dotted the grass and road, shiny with the slickness of rain.

Blustery blasts passed Ray as Dan and Ernest joined him on the small porch. Ivan's asthmatic coughs were all that remained as he struggled for air with which to torture them.

It was over. The damage was great but they had lived to see another day, and for that they gave thanks.

Michelle was nervous about getting on a horse again, but she would never express that to Des. She had not ridden one since finding out she was pregnant with the boys, eight years earlier. All desire to ride was gone, and Des had never pushed her to try again. He had his group of friends who shared his passion, so he rode with them.

Watching him saddle up his horse, she surveyed the others, looking for a likely candidate that would be gentle and tolerant of her nervousness and long-lost ability to ride. In fact, she was never a good rider. She knew how to hold the reins, but as far as letting the horse know who the boss was, it was quite obvious in the first few minutes - these beasts were, and they knew it!

"The one you will ride is big, but more quiet than the rest," Des said, as he began to place a halter over the head of a mammoth horse. "Just get on and keep the reins tight but loose, do you understand?" He cinched the saddle and the animal flinched in its sides. Michelle flinched too. "Climb up on the side of the truck there so it'll be easier for you to get on. Her name is Lucky Golden Nugget."

"This is the time that you pick to go riding?" A familiar voice called out. Des and Michelle turned in its direction and were greeted by the ever-smiling face of an old friend, Cat. She slowly pulled her car to the side of the road next to them. Cat was known for her off color comments and jokes, but seemed to be keeping a lower profile on this somber day.

Des pulled the horse next to Michelle, who was standing on the back bumper of the Explorer and hanging onto the luggage rack, waiting.

"Can't get through Seven Mile Beach by car, so we are gonna ride there," Des explained as he helped hoist Michelle atop the horse. She readjusted herself for comfort, then hesitantly took the reins from Des.

He busied himself with his own horse and saddle while continuing to chat with Cat.

"You'll know when that horse throws you off his back!" Cat mocked. "I betcha you won't wanna ride nothing again!"

This isn't helping, Michelle thought. Her animal was not keeping still. She swayed and moved and stepped every which way. Michelle tried to pull in the reins, but Lucky's head jerked at the slightest tug. "C'mon Lucky, be nice," Michelle begged beneath her breath.

Des was almost finished saddling his horse, a much smaller animal that looked a lot calmer. Michelle wondered how she had drawn the short straw on this.

"Well, just be careful, cowboy!" Cat called as she let her foot slide off the brake and slowly coasted away from Des and Michelle. "I'm going sightseeing." With that, she accelerated, causing the horses to startle and prance around all the more.

"Dessss…!" Michelle cried out nervously. She was losing control. Lucky was not happy, and at any moment threatened to buck Michelle right off of her back. She pulled the reins and the horse almost reared. "Get me down! Get me down!!" She screamed, and Des was there in an instant.

"We can't do this. They are too spooked. Just take me home and then you can go alone," Michelle whined, stepping away from the horse, and put the vehicle between it and her.

"All right," Des sighed. "Let me just put Lucky back and tie this one up. Then I'll take you home."

He quickly removed the saddle, which instantly relaxed the animal. Michelle watched Des deftly lead her back into the pasture with the other horses. She imagined that Lucky was recounting her ordeal to the others in her horse language.

TWENTY-FIVE

Pulling up to the T-junction, Des came to a halt out of habit and respect for the stop sign, not because there was any traffic to stop for. As he was about to accelerate into the turn that would take him home, his eye caught an approaching vehicle that looked vaguely familiar, even from this distance. He applied the brake and waited for it to get closer.

"It's Aunt Lyssa!" he exclaimed, his eyes widening with excitement as he slammed the car into 'park'. "She made it through to West Bay!"

Michelle turned to see the vehicle as it slowed to a stop on the other side of the road from them.

"Where you coming from?" Aunt Nay-Nay called from the passenger seat of her sister's car.

Des waved to her, his furious hand movement in the air threatening to stir up another hurricane right there. "I have been trying to get to you and Mummy!" he called out. "We were told the road was closed off. How did you get through?"

"We came through Governor's Harbour, but it is such a mess!" she called back. "Your mother is okay, but her house is gone...just gone!"

"Okay, we are going there now. Your house is

fine. I didn't go inside but everything looks good on the outside. We'll be back later." Des quickly put the Explorer in gear and shot out of the road.

The change of mood was so instant and so light. Though there was little conversation, the atmosphere was significantly brighter, knowing that his mother was safe.

He drove slowly, no longer pressed with the heavy need to find her and learn of her fate. They drove in silence past sand hills. So much sand had been moved from the beaches, as if a bulldozer had risen from the sea and plowed it across the road, into the bushes.

Corrugated zinc wrapped around trees, utility poles and columns. Roof shingles embedded themselves in the lightweight Marmaran siding of some homes. People wandered like zombies along the road, mouths agape, staring blindly ahead.

Des waved to each as he passed them, offering words of encouragement and blessings. Michelle uttered not a word. She rode in stunned silence, wondering what had become of this beautiful island paradise.

Monday afternoon brought a glimpse of the sun that had not been seen in two days. It peeked through the clouds in bright shards of light, streaking the grey sky in random bursts. The ground was soaked to the core, its earthen pores excreting puddles everywhere. The backyard was a patchwork of carpet grass and still water, which held perfect reflections of the gray-blue sky in its mirrored surface as if the heavens had fragmented and cast chunks of itself into the earth.

Mama DeDe's rubber-soled shoes squeaked and squished as she stepped onto the portions of grass that were seemingly devoid of water. Each step sank deeply into the ground that gave way beneath her weight, the soggy soil sucking in her footprint. She placed a box against the hedge that lined the farthest end of her yard, then stepped back to survey their progress. All day she and her fellow survivors had been removing items from the garage in the hopes that most of it could be salvaged. Her patio and garden were dotted with boxes, tools, toys, gadgets and other random items that she had either permanently stored in the garage or had temporarily housed there to ride out the storm.

"DeDe, what do you want me to do with this?" Lacy asked, as she held up a dripping tangle of clothes.

"Throw it on the pile by the side of the house near that washtub," DeDe answered, pointing in the intended direction. "After we get everything out, I want to try and wash up stuff like that so they don't get stained and ruined."

She wiped her sweaty brow with her sleeve and heaved a long, deep sigh. The humidity that trailed the hurricane was so oppressive it sucked the air out of her lungs. Her clothes were completely wet, a combination of the moisture on the items that she carried and the perspiration that escaped her. She walked through the French doors into the garage and grabbed a bottle of water to rehydrate.

HURRICANE IVAN: THE EXPERIENCE

Chloe sat on a beach mat on the floor, sorting the smaller contents of some boxes that had fallen victim to the monstrous tide the day before. Beatty lay asleep on the beach chair beside her, her prescription glasses slightly twisted on her face, one hand hanging limply off of the armrest. Mama DeDe gulped two more swallows of the warm bottle of water before reaching over to remove the glasses and right her mother's protruding arm. She patted Chloe on the top of her head as she turned to resume her tasks, snatching up another wet box full of something that tinkled inside, no doubt broken.

The men were upstairs, surveying the damage on the second level. Mama DeDe had not wanted to go up there just yet. She was still mulling over the images that she had last seen of her home. Snapshots played back and forth in her head, memories too fresh to be replaced with the inevitable destruction that certainly awaited her. The bomb site of her garage was enough sensory overkill for now. Let the rest be as it was until she could assimilate this reality, before she added more to it.

A loud thump and scrape punctuated her thoughts as Al and the others wrestled with something large and heavy up there. It wasn't a comforting sound. She exited the room into the backyard as Lacy stepped in. No words were exchanged, just looks of mutual exhaustion and sadness. Both knew the fate of Mama DeDe's home, but there was no word on Lacy's. Mama DeDe had convinced Lacy to stay with her throughout the duration of the hurricane until the 'all clear' was given, with the hopefulness that this would be just another passing storm. Now it looked like the visit would be a prolonged one.

Nay-Nay would also stay with Mama DeDe to help her recover as many of her possessions as possible, though she had insisted on getting to her own home as quickly as possible to ensure that it had survived. She lived in West Bay, next door to Mama DeDe's son, Des, and had caught the first ride that was going in that direction. Mama DeDe had not heard from her since.

The last communication with Des had been at the climax of the storm, when they were holed up in the closet, saying goodbye. Since then there had been no success with reaching anyone by phone, and this worried Mama DeDe greatly. She stood on the patio, under the eave that provided the faintest ribbon of shade from the peek-a-boo sun. No birds chirped or sang, no chickens clucked or crowed. No cars drove by on the other side of her hedge. The garden was eerily quiet. Not even a breath of wind stirred. Ivan had tromped through, snatching the birds, the sun and the joy out of her island life. He had robbed her of her home, her things, and her sense of security. She was overwhelmed and beaten, but just too strong a woman to let it reduce her to a sniveling heap of vulnerability and weakness. Mama DeDe had such an enormous job ahead of her that to stop and allow self-pity to take hold of what Ivan left behind would just defeat the whole point of being a survivor.

She squared her shoulders back, tipped her chin a little higher, and finished the walk to the end of her property to add the box to the pile. All was not lost. They still had their lives and their health. Each day would present its own challenges, but it would get easier with time. They would repair and rebuild and move on, and in time this would be a distant memory, the measure by which all other storms were compared.

As she turned away from the stack of damp boxes, the sun broke through a cloud and cast a golden stream of light around her. The carpet grass twinkled at her feet as the rays caught the puddles of water that were slowly evaporating in the heat of the day. Larger puddles glistened, radiating starry glitter that winked at her from weepy eyelets. The world was still spinning. This too would pass.

TWENTY-SIX

Along the road that winds from West Bay to the rest of the island, the Caribbean Sea surges mere feet away. The neck of the island is at its thinnest along the Seven Mile Beach, one of the most beautiful beaches in the world. On the other side of this strip of land is more water, though not boasting of any famous beaches. Million dollar homes and condos line the sea, vying for every square inch of paradise available. The view offered is comparable to none, an exotic panorama of every hue and tone of blue imaginable. Strangely, very few natives live on this side of the road.

Michelle marveled at how calm the sea was as she caught glimpses between the Seagrape trees and other foliage that withstood Ivan's incredible pounding. Well, not exactly calm. It still roiled and burped, spewing salt mists into the air as it fought to maintain its strength and claim more land. It just didn't appear as vicious as it had sounded just a few short hours earlier. It did not pack the punch it once had. She knew that it was dying, and wished for a swift death.

As she and Des left the western district, they

began to encounter their first real obstacles. For as far as the eye could see lay downed utility poles tangled with electrical and phone cables. The black and silvery snakes lay motionless in the road, slick with rain. Their struggle to remain hoisted midair above the traffic had been defeated by the toppling of their supportive wooden beams. Some poles lay splintered from apparent forceful separation at the exposed base, while others had been removed in their entirety from the ground, large, deep, gaping holes evidence of their extraction.

Des slowed the vehicle and carefully navigated over the wires, sure to avoid the poles. Traffic was heavier in this part, as people attempted to leave West Bay and others still tried to enter. Smiles and waves greeted each driver and passenger, as relief registered on seeing someone they knew.

The Explorer rocked back and forth as the tires alternated between road and cabling. The sound of rubber and metal gritting on the wet asphalt produced an unusual musical sound. Squeaking and scraping accompanied the engine's purr as Des pressed the gas to push the vehicle up and over each serpentine mound.

The condos that stood tall along the road varied from damage invisible to the passing motorists, to all-out demolition. Roofs were chipped and broken, windows smashed and blown out. Cars in the parking lots were buried either by sand or by debris.

Michelle sent up a silent prayer of thanks for her life, family and belongings being spared. Tears pricked her eyes as she thought of what was yet to be seen.

Ahead of them, traffic was backing up as drivers encountered what appeared to be a mountain of sand standing at least eight feet above ground. This was what they had been warned of. It blocked the road entirely and swept from the sea right to the front door of a major hotel.

Vehicles were forced to detour onto a back road, frequented by motorists who preferred to beat the rush hour traffic on their morning treks to work.

No left turn between 7am and 9am, read the partially fallen sign, as car after car ignored it and slowly rejoined the convoy ahead.

On the left lay the sprawling ruins of another hotel. Every window was boarded up, but all that remained of the roof were the wooden joists and beams that had once held and supported the zinc covering. The latter was strewn everywhere, and some pieces had even been halted mid-flight with such force that they curled around trees like foil.

The road led into the interior of the island through a residential area that boasted expensive homes nestled cozily on quays with small yachts and powerboats tied to each privately owned dock. Sunday drivers enjoyed this route, as it was flanked by giant Casuarinas that offered a pleasant covering from the persistent sun. Joggers and mothers pushing baby strollers also frequented the area, which stretched almost two miles.

As Des and Michelle rounded the corner that would take them along this favored track, confusion began to set in.

"Wait...you're not supposed to turn here...it's..." Michelle's voice trailed off, her eyes searching for a familiar landmark. "Isn't that...?" She couldn't figure out just where they were, even though she knew exactly where they were. For as far as the eye could see, those beautiful tall Casuarinas lay flattened across the landscape as if they had been used in a game of Pick-Up-Sticks. The bulldozer that had been Ivan had ploughed through the terrain with root-wrenching force, leaving not a single tree standing.

Michelle took in the view even as Des continued to follow the cars. Many had pulled over, and drivers and passengers were exiting their vehicles to survey the devastation. Michelle recalled the video clips of the effects of the atom bomb tests, where everything in its path for miles around was nuked within seconds of detonation. Only something so powerful could have caused this. Not a pathetic hurricane! Her eyes burned with wet emotion, slowly and painfully filling with one tear and then another until so many had backed up that they spilled over her lids and raced to her chin, first the right and then the left.

The cars ahead were weaving on and off the road to avoid fallen trees. The ground on the sides of the pavement was so saturated with rain and flooding that deep, muddy rivulets had formed. The tires of the Explorer squished through them, the engine roaring as Des was forced to accelerate to avoid getting stuck. Sections of the road were so impassable that some homeowners had pulled up their own hedges to create makeshift passageways through their front yards. The grateful drivers honked and waved as they passed through.

"Oh dear Lord..." Des whispered. "The boats!"

Michelle could barely tear her eyes from the barren vista of the treeless land on her right to see what Des was talking about. When she finally did, she wished she had not. The scene that lay on the left was beyond comprehension.

Between the fingers of land lay man-made inlets, which brought the sea to the back door of each property in this area. It was desirable property and many well-heeled residents took full advantage of the access to the open water. On any given day, the harbor would be dotted with a variety of watercraft ranging in price and size. Today was no exception, however, the boats were not slowly rocking beside their moorings.

As if frightened by the ravaging effects that the tumultuous sea threatened, each vessel had climbed onto land and was either snuggled against the home of its owner or pitched sideways in an open lot. The scene of damage and destruction was so overwhelming, it defied reason.

"Look at that one there," Michelle called out, as Des slowed to a stop, unconcerned about the people driving behind him. They in turn stopped to take in this incredible sight as well.

Michelle was pointing to a two-storied party catamaran that tilted precariously on its side, half of its underside resting on a wooden dock. Apart from its awkward angle, it did not appear to be damaged, though she doubted that it would be entertaining anyone in the near future. Behind it, and entirely on land, a large luxury fishing boat nestled between two homes. It sat perpendicular to the waterway, its cloth canopy shredded.

The discarded remains of a sailboat sat in an open lot, the tall mast jutting outwards, angling 45 degrees from the land. The rope that had once tethered it to a berth was slack on the ground stretching to the water. The boat reminded Michelle of her sons' toys, which were thrown down in the same manner when they tired of them.

Des pulled back onto the road and pressed on in silence. Row after row of quays offered the same view, only the style of home and boat changed slightly. The buildings around them bore no resemblance to what had stood there just the day before. Without the trees, nothing looked the same. Nothing would ever look the same.

Christy had enjoyed the best sleep she could remember in recent history. It was not the most comfortable to start with, but it was definitely uninterrupted. Rebecca had gone down after eight Sunday night and had slept right through until morning. No feedings; no crying. Ryan fell asleep shortly after that, having worn himself out playing with his newfound friend, Matt.

Christy had placed him on the twin bed and tucked him in before settling back onto the floor that had been heavily padded with numerous blankets and comforters. It was not as cushioned as a mattress, but it provided substantially more support than the lumpy one in her apartment that she had suffered on.

The group of new friends congregated in the kitchen and adjoining dining room over breakfast, as if they had been having sleepovers forever. The atmosphere was relaxed and lackadaisical. So completely different from the terror that she had experienced not so long ago. She wished she could stay here with them without end.

The hours whittled away just as the effects of the storm diminished to a regular storm. It still bore the bite of danger, but Ivan's tail could be seen tucked into the farthest glimpse of sky as he moved westward and northward towards the United States.

Christy was in no hurry to leave when it became obvious that they were rid of the worst of the hurricane. Her hosts did not force the issue either. Everyone knew that she had nothing to return to. In time, she would have to find somewhere to stay, she simply could not remain here, as there just was not enough room. But for now, she was content to remain at the invitation of Janet and Jim.

Something had blossomed out of the blackest night of the storm. A seedling was growing, heavily watered by Ivan, tended to by Christy and Keith. They were discovering each other in the most traumatic event of their lives. She knew not how this would proceed, nor did she spend much time obsessing over it. She was making a deliberate decision to let it grow at its own pace, as she observed the paradigm set down by the married couple who exhibited patience, love and tenderness.

She might have lost all of her worldly possessions in the hurricane, but it appeared that she had gained a valuable snapshot of a world of stability and constancy. She liked it very much. There was also the growing possibility that this blossom that she shared with Keith would flourish into something real and true.

As she looked out of the window into the calming storm, a pinprick of light escaped the layers of cloud that had enshrouded her world. The bright beam shot to the ground. The symbolic nature of this sight was not lost on her. She saw her own sunbeam in her personal life, and happily pushed away the clouds that had enshrouded her for much of her life.

TWENTY-SEVEN

On the main road once again, Des surveyed the buildings all around. Some of them looked quite fine, a few roof tiles missing maybe, but generally in good condition. Perhaps there was hope that the devastation was only in certain areas.

The road he drove on now was called Seven Mile Beach or West Bay Road. It connected his district to downtown by way of the strip, which had become home to the island's largest concentration of tourists. Hotels and condos lined the beach and restaurants, bars and nightclubs with bright neon lighting usually beckoned from the other side of the road. Traffic was normally thick and heavy along this stretch, but not today. Vehicles were certainly traveling through here, but in significantly diminished numbers. Everyone was still trying to comprehend the damage to his or her own property before venturing out to gawk at everyone else's.

A knot in Des's chest was rising up his throat as he began to imagine what his mother's home had been reduced to. Her road loomed up ahead, light poles and shingles scattered about causing him to weave and swerve until he reached the turn.

The first house at the top of the road must have been a tasty morsel for Ivan's hunger. The second story lay open to the world, all contents visible. In this subdivision, as with everywhere else they had seen, the trees were either completely uprooted and washed away, or were stripped bare of all leaves. What did remain standing projected bony finger into the swollen, gray sky, pointing accusatorily at the cause of their nakedness.

Des took each turn carefully and slowly, easing closer to his mother's home. He now knew that she was fine, but that her home had not survived the thrashing of the storm. Nay-Nay did not go into much detail about the damage, but Des's recollection of the frantic phone call and the sad look on her face evidenced a sad conclusion.

Now he was on the straightaway, with only five more houses until he reached hers. It was the only one with a high Ficus hedge surrounding the property and boasting an arch over the entrance. Mama DeDe had taken pride in training the tree to grow that way. The hedge was thick and green, perhaps three feet deep but no less than two. It provided tremendous privacy and was attractive to look at as well. On the inner side of the hedge, a tall chain-link fence added additional protection. A six-foot gate on wheels was usually drawn across the entrance when Mama DeDe was not at home or had retired for the evening.

As Des and Michelle neared the house, the first and most obvious thing they noticed was the absence of the majority of the roof. Only one side of the house remained covered, and even that piece was badly damaged and caved in. The other side was totally bare, completely devoid of even a single rafter or truss. It was as if a giant hand had come down and simply plucked the lid of the house off like a Lego, unhappy with the look of what had been constructed. Where the roof was now was not apparent. Though smaller pieces littered the yard, the enormous triangular structure was nowhere to be seen.

They pulled up to the driveway and immediately noticed that the Ficus archway was nothing but a knotty mess of broken limbs. The hedge was badly mangled and twisted, and gaping holes allowed unwanted eyes to peer in and survey the damage without entering the property. The gate was pulled across and secured with a heavy chain and padlock. A plastic sign was tied to the small metal wires of the chain-link advising visitors and intruders alike to enter at their own risk: *WARNING: Bad Dog*, it read, with the ugly mug of a pit bull whose collar bore large spikes and teeth lay bare with a snarl. Des fished out his keys and unlocked the padlock

The sound of the wheels rolling across the gravel alerted the occupants of the house, and a voice from inside the garage called out, "Who is it?"

"It's me, Mummy," Des called back, relief pulling the weights of worry off of his shoulders and down his back. He left the Explorer in front of the driveway and beckoned for Michelle to join him. She shut off the engine and exited the vehicle.

They both instinctively approached the set of garage doors, but realized that they would not budge without electricity.

"Come 'round the back, honey," Mama DeDe called out. "Those doors are stuck."

Together they walked around the side of the house, stepping over heaps of junk that seemed to have been accumulated into one spot at the same time. It seemed odd that so much had been pressed up against the house like that and just stayed there during the gales. The yard was largely unrecognizable. If Des had been blindfolded and brought here, he would not have been able to identify where he was. Nothing was where it should be.

The backyard was the same and worse. Added to the chaos of rubble and wreckage were the entire contents of the garage, arranged in an organized mess. Lacy and Chloe were hanging towels on a clothesline that had obviously been erected recently. Des and Michelle found Mama DeDe inside, seated on a footstool and sorting through a soggy cardboard box. Michelle instantly ran to her mother-in-law and embraced her.

"I am so glad you are okay, we were so worried," Michelle sobbed, straightening herself and wiping her eyes.

"Child, I am right here!" Mama DeDe laughed, amused at Michelle's reaction and perhaps a little uncomfortable with the attention. "We are fine, but my poor house is gone."

"Mummy, as long as you are alive, that's all we need to worry about. This house is just a thing, a possession. It can be fixed. Your health and life are the main priority and that's all there is to it," Des stated matter-of-factly. He bent down to kiss his mother's cheek and she, in turn, kissed his.

"Yeah, well how is the house in West Bay? Did you get much damage? How are the horses? Oh but first, how are the boys? Did you bring them?" Mama DeDe asked, looking around to see if they would appear around the corner.

"They're fine, Jackie has them," Des said, and sat down in a plastic garden chair as he began to tell his mother of his personal Ivan experience.

EPILOGUE

In the weeks that followed the hurricane, Michelle learned that the luxuries of life were not really the essentials. Essentially one only needed a handful of things: water; that was a scientific fact, for without it the human body would die. Food: again, scientifically proven to be a life-sustaining fundamental, and air: oxygen ultimately keeps the heart beating. Everything else, from ice to gasoline, was just a benefit to one's existence. Of course, a modern day human feels that in order to thrive happily, one must have the host of accoutrements that society (and ingenious marketing) determines are an absolute must in order to survive this world.

Fortunately for Michelle, she had access to the three main necessities and was fast learning to do without the social ones. She had not had an ice cold drink since the day the electricity had gone out and the bags of ice had melted in the Igloo. She had not slept in air-conditioning, nor had she had a hot shower, or for that matter a shower that actually flowed from a showerhead. Des had devised a system of bathing that only involved a five-gallon bucket, a cup and a candle. Very romantic, but very chilly even in the muggy heat that Ivan left behind when he finally evacuated the islands in search of more challenging game.

Michelle was introduced to the laborious task of washing every piece of clothing by hand in a large metal tub outside. She initially missed the softness that the dryer sheets gave a fluffy towel, but soon learned to appreciate the crispness and fresh smell of sun-dried sheets. Clotheslines stretched from every possible pole, tree and fixture that remained intact.

Cooking was something that did not require much sacrifice, as their full-sized gas stove was obviously not reliant on electricity. Within days, they were forced to cook everything in the deep freezer to ensure that nothing that thawed would spoil. With so much food to cook in such a short period of time, the only reasonable thing to do was share it with the neighbors. Everyone enjoyed lobster, shrimp, chicken and beef until everything was completely consumed. The buffet lasted less than a week, but guaranteed that there was no wastage and their strength was fortified for the days ahead, where a hard day's work took on a new meaning.

Michelle's stepmother had called days before the hurricane's landfall to inform them that the water companies would be turning off the water supply well before the strike. She advised them to fill up every available container with water, just in case there was a delay in reconnection after the storm, including the bathtub. Wisely taking her advice, Michelle asked Des to find something large enough. "I've got just the thing, or rather, things," he had announced, and returned with two fifty-five gallon drums, which he thoroughly disinfected and filled with fresh city water.

It was this supply of water that kept them going when no other source was available. Eventually, one of the water companies would deploy trucks to each district, pumping fresh water into containers for the multitudes that would gather. Announcements were made by radio or word of mouth, and the lines were long in the baking sun. People queued for anything that was being handed out - water, roof tarps or gas. It mattered not. If it were free, you would have a line.

Of the three major supermarkets on Grand Cayman, only one was able to reorganize and reopen its doors right away, though it was a very limited provision. The owners were able to convert the adjoining warehouse into a crude market, and placed security and armed guards outside to mitigate potential rioting and unruliness. Items that had not been touched during the flooding or affected by the collapsed roof in the main shopping areas were collected and restocked on shelves in this enormous storehouse. Entry was controlled and limited, only a select number of people were allowed inside at any one time, which made the wait long and exhausting. Industrial-sized generators ensured that the wait was worth the discomfort, as a fresh, cool blast of air greeted each person who walked through the doors. Enormous fans blew onto the shoppers throughout the store, and provided a brief interlude from the scorching heat.

Thankfully, Michelle, Des and the kids had no need of the stores right away, as their personal shelves were substantially full of canned goods and dry snacks. Forward thinking allowed them to avoid queuing up in the sun for hours on end, and focus on the arduous undertaking of removing the debris, dead trees and animals from their property. Yet on a day that took them sightseeing, they decided to have a look inside, on the off chance that there was something of interest for them in there.

The line of people snaked from the door around in a zigzag pattern to the parking lot. People pushed and inched forward to move towards the walls where they could at least enjoy the shade. Mumblings and mutterings suggested many were not happy with the wait, but were too fearful of the security to speak their minds openly.

The foursome joined the end of the line and waited for their turn to enter the Mecca, which promised a bountiful selection of goodies that were now highly coveted. Time drew on, each step towards the door divided by a generous multiplication of time. Feet shuffled forward, two steps; stop, three steps; stop. The children grew agitated, but Michelle calmed them with promises of a treat that surely awaited them on the inside. She felt as if she were waiting for a magical ride at a theme park, where each new adventure first required patience and long-suffering. The true enjoyment was realized only after the angst of being subjected to the interminable wait. She hoped that this would be the case here, too.

The merciless sun burned their necks, while simultaneously slowly roasting the pavement of the parking lot, the acrid smell of tar rising up and repressing the oxygen in the air. Heat shimmied in hazy walls, creating mirages of water in the distance. The sky above was a brilliant blue and utterly cloudless, belying the true potential of what could be contained and crafted between the heavens and the earth. The day was picture perfect and gorgeous.

Finally they were next in line to enter and the anticipation bubbled within. How could anyone get so excited about entering a mere grocery store? What a crazy turn of events that would sharply bring all things into clear perspective. A little over a week ago, this would have been a mundane experience, barely worth mention; and now it was the highlight of their day.

The guard raised his arm, which had been blocking the doorway, a signal that it was permissible for them to enter. Michelle put her hands on the backs of her sons and gently pushed them inside, Des walked directly behind her. The coolness of the inside welcomed them, their eyes darting in every direction to take everything in. The room was two stories in height and filled with produce and dry goods. Standard metal store shelves were arranged to form aisles, and shoppers were busily passing around each other to grab as much as their shopping carts would hold. Freezers were arranged at the outer perimeter with whole frozen chickens, while baskets on the floor were filled with batteries and flashlights.

Michelle took a shopping cart and proceeded down the first aisle with Des and the children in tow. She was very selective, ensuring that her choices were justified and necessary. From all accounts, the island would be in this condition for a long time to come and she needed to be extremely frugal, making every dollar count. Down the first row and up the second she slowly walked, eyeing everything on each shelf as if she had not seen that item before in her life.

Dylan was the first to spot the tall refrigerator at the end of the second aisle. It was cleverly set in front of a cashier and angled just enough to catch the attention of everyone who walked down this row. Its glass door was partially frosted and its contents beckoned. Inside was an assortment of drinks, milk, juices, water, all chilled and ready to consume. He tugged his mother's sleeve and pointed, looking at her with sincere appeal. Demitri's attention was also drawn to the glass casing, and he immediately begged for one of the beverages.

Michelle pushed the cart towards it and stopped. She theatrically put her hand on the door handle and paused for effect. "Aw Mommy, c'mon, you're killing us here!" Dylan objected, "Just open it for goodness sake!" He tried to grab the door, but she pressed it harder against the unit.

"Mommmmeeeeee!" Demitri protested, "I want a drink." Michelle laughed and threw the door open, the cold air wafting against their skin and producing instant goose bumps. She glanced over each shelf, unable to decide on what to take. Des grabbed a small carton of milk and quickly tore open the spout. He lifted it unceremoniously to his lips and poured the contents into his mouth, causing a thin line of white liquid to slide down his chin. He gulped and gulped, relishing the coldness. The other three stood there watching him, their own desire to drink suspended as they gawked at his unadulterated satisfaction of thirst.

Finally he withdrew the carton and gasped for air, his moustache dripping with milk. He grinned, totally satiated. Des passed the carton to Dylan who finished it off in like fashion. Michelle reached inside for another carton and handed it to Demitri, then retrieved one for herself. The four of them stood in front of the refrigerator, heads tilted back, guzzling the first cold drinks that they had enjoyed in weeks. The open door continued to fan them with frigid air.

Energized and refreshed, they finished their shopping and returned home with more cold drinks, which they consumed before nightfall.

"You know what I didn't realize before?" Des asked as he stared up into the night sky. "I didn't realize how many stars are in the sky." He sat on the front porch of their home with Michelle and the boys. This had become their nightly ritual and was a very effective wakeup call for how little they had previously shared each other's company pre-Ivan.

With no power, there was no television, no games, and no distractions. The setting sun signaled not only that the day had come to an end, but also that each person was required to head for home. The government-imposed curfew was strict and heavily enforced. Anyone seen on the road after dark was stopped and questioned. Looters had become as thick as the mosquitoes, and many abandoned and ruined homes were being targeted. Measures had to be put in place to eliminate the threat.

Neither Des nor Michelle was bothered by this imposition. Whatever could not be achieved in daylight would simply wait for the following day to come. They had nowhere to be when the sun went down, other than right here with each other and their children.

Above, the black sky exposed tiny diamonds that twinkled and winked. Some were white while others flickered green and red and blue. With no street lamps and no illumination from the bigger towns that usually lit up the night sky with their heavy concentration of bright lights, the only place to look was up. All around them, the pervading darkness was absolute. The heavens provided the only luminosity and the show was beautiful in its simplicity.

The family stretched out in a single hammock, parents balancing the weight by sitting knee-to-knee with the boys planted in the middle, their heads each resting on an adult's chest. Only Des' feet actually reached the floor, so he was able to rock the hammock slowly.

"Don't ask me to tell you how many there are supposed to be," Michelle answered. "I don't think anyone *really* knows, even if scientists claim to have a figure."

"Tell us about when you were a kid again, Daddy," Dylan begged. "When your grandmother used to swing you in the hammock and get you to sleep." Shortly after the realization came that the island was going to take months to recover, and that electricity and water were not in the immediate future, Des had begun to fill the night hours with stories of his childhood. A time when there were no electronic games and not every household owned a television set. The children were mesmerized with how well he survived this unfortunate upbringing. How he coped without those necessities amazed them.

He filled their imaginations with tales of his boyhood adventures, playing outside with his friends and going fishing. With wide eyes and mouths agape, Demitri and Dylan listened with rapt attention. Neither uttered a word except an occasional "wow" or "how did you do that" when something he said excited or confused them.

The citronella candle burned on the floor near Des' foot, its pleasant odor filling the porch and chasing away the mosquitoes. Michelle watched the lazy flame dancing on the wick, the reflection doubling in the crystal depths of her eyes as Des looked from her to each of the boys.

Des relished this night more than the others. Tomorrow, the boys would be sent to the U.S. to stay with their uncle, and he did not know when they would return. He and Michelle had made the painful decision a few days earlier to send them away. He knew that it was the right thing to do, but it ripped his heart to separate from them. Grand Cayman was not the place for them now. The school they attended was completely destroyed and no one knew when it would ever be fixed again. While their food supply was fine for now and the supply of fresh drinking water was no longer a concern, Des felt that it was not a healthy environment for two seven-year-old boys.

Michelle's brother, JD, had agreed to take them and put them in school in Texas. Michelle had also decided to send their nanny, Gela, so he was comforted that between the two of them, they would certainly be able to care for the boys much better than Michelle or Des could in these circumstances. It was for the best.

Des forced the thoughts of tomorrow from his mind and continued on with his stories. The boys slowly closed their eyes and drifted into slumber, where dreams of another time and place circled in their minds.

CPSIA information can be obtained at www.ICGtesting.com
Printed in the USA
LVOW12s0158311014

411369LV00001B/8/P